CO

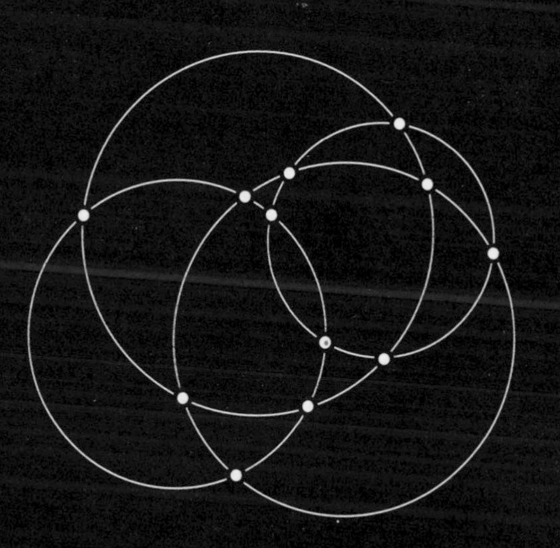

The MIT Press Essential Knowledge Series

A complete list of the titles in this series appears at the back of this book.

COLLABORATIVE SOCIETY

DARIUSZ JEMIELNIAK AND
ALEKSANDRA PRZEGALINSKA

The MIT Press | Cambridge, Massachusetts | London, England

This book was set in Chaparral Pro by Toppan Best-set Premedia Limited. Printed and bound in the United States of America.

Library of Congress Cataloging-in-Publication Data

Names: Jemielniak, Dariusz, author. | Przegalinska, Aleksandra, author.
Title: Collaborative society / Dariusz Jemielniak and Aleksandra Przegalinska.
Description: Cambridge, MA : MIT Press, [2019] | Series: MIT Press essential knowledge series | Includes bibliographical references and index.
Identifiers: LCCN 2019008960 | ISBN 9780262537919 (pbk. : alk. paper)
Subjects: LCSH: Online social networks. | Social networks. | Cooperation.
Classification: LCC HM742 .J46 2019 | DDC 302.30285—dc23 LC record available at https://lccn.loc.gov/2019008960

10 9 8 7 6 5 4 3 2 1

CONTENTS

SERIES FOREWORD

The MIT Press Essential Knowledge series offers accessible, concise, beautifully produced pocket-size books on topics of current interest. Written by leading thinkers, the books in this series deliver expert overviews of subjects that range from the cultural and the historical to the scientific and the technical.

In today's era of instant information gratification, we have ready access to opinions, rationalizations, and superficial descriptions. Much harder to come by is the foundational knowledge that informs a principled understanding of the world. Essential Knowledge books fill that need. Synthesizing specialized subject matter for nonspecialists and engaging critical topics through fundamentals, each of these compact volumes offers readers a point of access to complex ideas.

Bruce Tidor
Professor of Biological Engineering and Computer Science
Massachusetts Institute of Technology

ACKNOWLEDGMENTS

We are deeply grateful to Vasilis Kostakis, who gave us a great deal of constructive feedback on the early drafts of many of the chapters, providing thoughtful advice, suggestions, and comments. Lots of great ideas came from Gabriel Mugar, who not only helped us with refining some arguments but also directed us to some interesting angles. Marcin Zaród gave us much useful feedback, and we are very thankful for his support and help. Also, we wish to warmly thank Agata Stasik, Karolina Mikolajewska-Zajac, and Armin Beverungen for their insightful responses to some of the chapters. We are grateful to Mikolaj Golubiewski for his language corrections and editing suggestions.

We would like to express our gratitude to Editorial Director Gita Devi Manaktala at the MIT Press for her continuous support throughout the process. The number of excellent comments, corrections, and radical improvements made by Mary Bagg, the copyeditor the MIT Press assigned to our book, has put us in a weird limbo of gratitude, amazement, and awe.

Bruce Petschek, who was a landlord for both of us, deserves credit for providing us generous support and a friendly atmosphere, which made this book possible, as it did for many other books by other authors who stayed with him before.

We decided that while working on this book it would be only natural to rely on collaborative editing software. All chapters were written in Google Docs, with references managed through a Paperpile plugin, which is definitely worth the cost of its subscription. This approach allowed a fully seamless cowriting and coediting experience, as well as making it much easier to simultaneously request feedback from several colleagues and experts in the field.

In a truly collaborative spirit, we worked on all chapters together. Dariusz took the lead in the introduction, however, and in the chapters "Neither 'Sharing' nor 'Economy,'" "Peer Production," "Collaborative Media Production and Consumption," and "Collaborative Knowledge Creation," while Aleksandra spearheaded "Collaborative Social Activism and Hacktivism," "Collaborative Gadgets," "Being Together Online," and "Controversies and the Future of Collaborative Society."

Writing this book was possible thanks to Dariusz Jemielniak's grant from the National Agency for Academic Exchange (PPN/BEK/2018/1/00009) and supported by Aleksandra Przegalinska's Polish Ministry of Science and Higher Education's Grant Mobility Plus (DN/MOB/102/IV/2015).

Dariusz dedicates this book to his wife, Natalia Banasik-Jemielniak, whose continuous support, encouragement, and love made working on this project possible. Aleksandra is extremely grateful for the rock-solid backing and understanding of her passion from her husband, Jedrek Skierkowski, and for the constant support and encouragement of her mother, Zofia Przegalinska.

1

INTRODUCTION

A stranger on the street asks you for directions. Do you ignore the request or stop to reply? Most of us, unless we're really in a hurry, will stop to help even though we don't "get anything" for our effort. We react this way because we are extremely social animals—so much so, in fact, that social isolation can elicit distress exceeding physical pain.[1] Also, as humans, we are wired for collaboration: even 14-month-old infants are inclined to altruistically help others achieve individual goals, and are also willing to cooperate toward a shared goal.[2] Being collaborative distinguishes us as human even more than our opposable thumbs; the drive to cooperate significantly sets us apart from chimpanzees, our closest cousins in terms of DNA similarity, and this difference is already visible in young children.[3,4] In fact, cooperation with nonkin, so typical for humans, is relatively rare in the animal kingdom.

The drive to cooperate significantly sets us apart from chimpanzees, our closest cousins in terms of DNA similarity, and this difference is already visible in young children.

Collaboration may be an evolutionary mechanism that emerged as a result of the way our early ancestors acquired cooperative practices regarding food.[5] Sharing resources seems natural to us; indeed, much human sharing follows strong social norms of equity and fairness.[6] As Matthew D. Lieberman shows, even our brains react in deeply social and cooperation-oriented ways.[7] We are, in fact, wired to be super-cooperators.[8] *Homo sapiens* may have evolved via selection for prosociality in a process Brian Hare calls the "survival of the friendliest."[9] This quite powerful drive to cooperate with others starts early in life; studies show that the preference for prosocial behavior in humans begins as early as infancy.[10]

The desire to help may be intrinsically motivated in children as young as two and three years old, although, counterintuitive as it seems, extrinsic rewards may undermine that desire.[11] Even so, some research suggests that infants' prosocial behavior is actually governed by basic cost-benefit analyses.[12] This is not to say that as a species our cooperative behaviors necessarily derive from altruism. Rather, they may stem from mutualistic collaboration, a necessary result of interdependence within a group, which allows for skills specialization.[13] One way or another, though, collaborative behaviors define quite clearly (and to a large extent) what makes us human.

But human collaboration is a complex phenomenon. A great deal of capitalistic and neoliberal philosophy relies on positioning people as individualist, and mainly *Homo economicus*, rather than as being interested in participatory culture.[14] Perhaps this is why we've only recently become more attuned to the rise of collaboration—enabled, as it is now, by the development of communication technologies largely independent of traditional organizational structures, procedures, institutions, and hierarchies. This is particularly true when new tech does not impair sociability by supplanting our deeper offline engagement with superficial online engagement, but makes deeper engagement possible in the first place, or complements our existing offline connections.[15]

Human collaboration enabled by technology can occur in specific contexts such as peer production and collaborative consumption, or more generally through online sharing and exchange platforms. Emerging technologies, thanks to their direct collaboration-enabling features and their engagement of much broader populations, act as super-multipliers for many effects of collaboration that would otherwise be less noticeable. We perceive this phenomenon to emanate from what we call *collaborative society*: an emerging trend that changes the social, cultural, and economic fabric of human organization through technology-fostered cooperative behaviors and interactions.

This budding collaborative society relies on different modes of cooperation, sharing, joint creation, production,

Emerging technologies, thanks to their direct collaboration-enabling features and engagement of much broader populations, act as super-multipliers for many effects of collaboration that would otherwise be less noticeable.

distribution, trade, and consumption of goods and services by people, communities, and organizations. Collaborative society can also be viewed as a series of services and startups that enable peer-to-peer exchanges and interactions through technology. Although it may be a relatively recent new-technology-enabled phenomenon, collaborative society in its entirety is a system with good old sharing and collaboration at its heart.

In fact, cooperation laid the foundation for the internet as we know it today. To the developers of ARPANET, sharing knowledge proved more productive than trading information.[16] "Sharing" was also understood literally, as the early computer systems were premised on the collaborative use of hardware, and the culture of sending demos and software on floppy drives by mail long preceded online piracy.[17] The "hacker ethic," sparked by joining forces in imaginative ways, foregrounded radical openness, decentralization, and collaboration.[18] Most tech developers understood collaboration and peer production broadly, as the heart of the net's innovative capacities, and viewed encapsulated proprietary networks as limiting them.[19]

Marcus Felson and Joe L. Spaeth, in their study "Community Structure and Collaborative Consumption," introduced the term *economy of sharing* in 1978.[20] Today, more than 40 years later, the economy and society continue to rapidly develop and transform. The significant boost in sharing, as both a work- and lifestyle, stemmed from swift improvements in technologies such as social media, mobile technologies, cybercurrencies, and coproduction. Some experts, however, predict that this development will soon slow down, even though others argue that it has hardly begun.

Juliet Schor observes that a clear definition of technology-mediated open collaboration might be difficult to pin down.[21] Some already acknowledge sharing as an important term of self-description for digital cultures.[22] But others see the very concepts of open collaboration or sharing economy as still unclear and problematic because they embody at least a half dozen diverse phenomena: acts of communication, gifting, swapping, distributing, contributing, and digital replication.[23,24] Some scholars note that sharing is no longer just a mundane practice but rather "an expression of a utopian imaginary ... associated with politics, with socialist, communist, and anarchist values, with the free culture movement and the digital commons."[25]

Notions of "sharing" and "collaboration," however, proliferate in popular discourse and among decision-makers. For instance, Italy passed a bill in 2016 that defines the sharing economy as an "economic system generated by the optimization and shared allocation of space, time, goods, and services through digital platforms."[26] Similarly, the EU Commission defined collaborative economy as a brokerage between individual service providers and recipients.[27]

Even more important in the general discourse are the plethora of authors who believe that collaborative economy—particularly commons-based peer production—has the potential to transform the capitalistic system and the whole of society into something much less corporate driven and more equality based.[28,29] Proponents of open collaboration argue that online connectivity allows provisioning based on access rather than ownership, which enables better use of previously underutilized assets like cars, private rooms, apartments, and tools.[30,31] The thriving marketplace that exists as a result, they say, transfers the burden of maintenance to the provider while the sharing platform reaps the benefits.

At the same time, however, stark critics of the phenomenon insist that collaborative communities based on sharing may be yet another cover for social injustice and user exploitation.[32,33] With the advent of highly successful corporations that have embraced the collaborative rhetoric, such as Facebook, Uber, or Amazon's Mechanical Turk, the early promises—that internet users would be empowered through their participation in creating the online world—seem evermore distant considering the ubiquitous online presence of these companies and their influence on users' lives. For some companies attempting to benefit from the "sharing economy," the notion of sharing has become a marketing strategy, thus redefining sharing in the eyes of the millions who partake in these practices. Two examples of "sharing economy" fallout illustrate this point: First, unionized taxi drivers recently protested that apps such as Uber allow almost anyone with a driver's license to become a de facto taxi driver, but legal challenges in some jurisdictions have established that these people, whom the platform portrays as independent "sharers," are actually Uber employees. Second, lobbyists in Airbnb-infiltrated cities like New York City and Berlin have begun to appeal to regulatory and zoning boards to ban short-term rental platforms, often to the joy of the local communities whose rental markets suffer from shortages and inflated prices.

Although we believe these discussions to be important, we must point out that the rise of collaborative society extends beyond the phenomenon of the sharing economy or the deconstruction of capitalism. Yes, economic transactions play a role in technology-enabled cooperation. But collaborative society further encompasses the emerging fields of remix culture, peer production, gift economy, citizen science, collaborative media consumption, wearables tracking, and digital communication—all of which alter the very fabric of our societies but not necessarily (or always) by their transactional character.

Looking at this composite picture through a single lens allows us to see the larger, more coherent change that is taking place. Nicholas John compellingly argues that we are entering the age of sharing.[34] We mostly agree, but we believe that the principal weight of the change lies elsewhere. Whereas sharing is an important symptom of the transformation that substantially alters economies and markets, the new modes of technology-enabled interactions and their cooperative character are what make those changes so profound. It is not only the result of the collaboration that matters, but the collective action itself that affects how we grow as a society. Different platforms and business models may come and go, but the collaborative society—with all its monetary and nonmonetary aspects—is incessantly on the rise.

Once we realize that Airbnb is but a platform for apartment rental, we'll be more open to notice where else the collaborative society revolution strikes, and with much greater impact: consider Wikipedia, The Pirate Bay, 4chan, Tinder, and Twitter, for instance. While only moderately affecting trade, these repositories redefine not only the traditional knowledge hierarchy, but also our trust in intimacy, matchmaking, and friendship, as well as in public and private communication. Clearly, the consequences of the open collaboration phenomenon surpass recent reforms in capitalism. Although we describe various effects the collaborative economy has had on all our lives, we focus in this book on important social aspects of collaborative society. We are entering an age of networked individualism fostered by perpetual online connectedness.[35] It relies on a highly cooperative collective, irrespective of any financial dimension.

Unsurprisingly, the future development of collaborative society likely depends on the performance of Big Data analytics, the use of machine learning, and the development of the Internet of Things. There remain, however, other significant factors to consider, including the future of employment in the context of ongoing automatization (and the role of the state in forthcoming regulations), as well as the growing (or declining) willingness of people to collaborate, help others, and share their values and resources.

With this book we aspire to present a balanced, unbiased view of the collaborative aspects of new technologies: not just another take on collaborative society embedded only in critical theory but rather a forum to discuss potential future scenarios regarding open collaboration as a social phenomenon. With that in mind we showcase emerging practices of collaboration as they've occurred in individual communities.

To help ground our readers in our approach to these cases, we define the concept of collaborative society that we adhere to throughout the book: *an increasingly recurring phenomenon of emergent and enduring cooperative groups, whose members have developed particular patterns of relationships through technology-mediated cooperation.* Although we don't dismiss the major influence collaboration has on capitalism, we prefer to emphasize its impact on other areas, such as culture, intimacy, and relationships. To that end, we organize our book in the following way.

In chapter 2, "Neither 'Sharing' nor 'Economy,'" we discuss the key concepts of sharing economy, collaborative economy, platform capitalism, cooperativism, and gig economy. We criticize the persistence of "sharing economy" as a term in popular and academic discourse by critically analyzing its oversimplistic and fuzzy meaning. We also point to the rhetorical hijacking of the term by many corporations. Although we recognize the explanatory value of the other concepts, we furthermore demonstrate that many aspects of the ongoing major change in society do not necessarily relate predominantly to the economy. Hence, we propose the concept of collaborative society as a better way to capture the ongoing transformation.

With chapter 3, "Peer Production," we present both the narrower view, which perceives peer production as principally commons based, and the wider perspective that includes commercial initiatives. We describe the ways in which peer production operates, such as its peculiar approach to authority and leadership, its heterarchical structure, and the open participation model. We observe the consequences of consensus-orientation and participative decision-making for conflicts, gender balance, and formalization. In addition, we discuss the motivations to participate in peer production, which may include a person's desire to create a portfolio, a professional reputation, and a standing within the community, as well as a desire to do something for the general good and to collaborate with strangers. Then, as we turn to address what happens when producers and consumers of collaborative media become one and the same, or *prosumers*, we look at the internet as both "playground and factory" with the potential for exploitation through unpaid work gamified as fun.

In the fourth chapter, "Collaborative Media Consumption and Production," we discuss online cooperative initiatives that do not necessarily focus on long-term production of larger works but rather foster the joint production and consumption of media. We then explore the consequences that the mass participation of "amateur" creators in collaborative media may impose on our traditional perceptions of "professionals," revealing that in some industries consumers may actually prefer amateur content, while in others the criteria of what is amateur and professional are definitely not set in stone. Moreover, we explain how collaborative entertainment consumption and production rely on a performative value of collective participation, rather than on creating a lasting outcome. In our discussion of memes, we focus both on their comic and social activist manifestations.

We consider how new platforms of communication as well as content creation and redistribution have stimulated the birth of new social movements in chapter 5, "Collaborative Social Activism and Hacktivism." We discuss the culture of hacktivism and the blend of technological and social focus in its collaborative initiatives. Examples of the Anonymous and anti-ACTA movements show in detail how cyberactivism originated and continues. Finally, we address the criticism of slacktivism and consider the political cyberterrorism aspects of hacking.

In our sixth chapter, "Collaborative Knowledge Creation," we begin with a thorough exploration of how trust in general science, and authority of academia itself, has deteriorated. We furthermore consider the phenomenon of citizen science communities as an emerging new model of collaborative knowledge creation. Then, we discuss their relation to do-it-yourself (DIY) science and biohacking: new communities working outside the traditional methods and environments of scientific research but no less determined to make actual discoveries, and to liberate scientific discoveries from the hands of licensed academics. Ultimately, we contrast these communities with alterscience movements, which aim to question the existing knowledge hierarchies by offering knowledge systems based on the beliefs of their members.

Continuing the discussion about participative and spontaneous participation in DIY movements and self-tracking in chapter 7, "Collaborative Gadgets," we describe the various uses of widely available tracking technologies. We examine the principles of the Quantified Self movement and position it within the collaborative society approach. Then we look at various gamification models that rely on wearable trackers and discuss how they affect social behavior. Finally, we refer to the emerging electroencephalography (EEG) self-tracking communities and discuss the whole concept of collaborative tracking as an idea that could one day replace self-tracking.

In our eighth chapter, "Being Together Online," we consider how the new, internet-intermediated forms of social relations reshape the ways in which people interact. Using the example of the online virtual world accessible in *Second Life*, we show how fully immersive online games from the past enabled collaboration. We then turn our discussion to how users reposition online platforms for media sharing or dating according to the needs of collaborative society, looking specifically at Instagram, Snapchat, and Tinder tools for cooperation. Thus, we show that collaboration is a default human orientation, further enabled and amplified by online platforms.

In our ninth and final chapter, "Controversies and the Future of Collaborative Society," we extrapolate from the current trends and observe that collaborative tendencies may not necessarily prevail. We discuss the negative effect of collaborative society's advances, including filter bubbling and fake news. We remark on the dangers and potential of bots and other nonhuman actors that now enter the collaborative society, but we also discuss being open to their positive capabilities. Finally, we consider the possible impact of intermediating technologies on the future of collaboration.

2

NEITHER "SHARING" NOR "ECONOMY"

Open collaboration, *sharing economy*, *platform capitalism*, and *peer production* all describe certain aspects of a revolutionary change resulting from sociotechnological advancement. In this chapter we briefly describe these concepts and explain why we believe that viewing them through the lens of the collaborative society better suits aspects of their transforming potential, which could otherwise go unnoticed.

Open collaboration is a form of organization and cooperation in which participants share a common goal but are loosely coordinated, yet together they create a product or service and make the final result available to anyone interested.[1] Open collaboration communities often rely on common discussions, but interactions among their participants are usually optional and depend on the depth and complexity of the collaboration. The idea, which originated in free/open source development circles, is often linked to the GNU Manifesto.[2] Written by Richard Stallman, the document postulated the creation of a Unix-compatible operating system that would be available, free of charge, to use, study, change, and redistribute.[3] The idea worked well, and not just for an operating system: it currently reaches beyond programming—one example being in the open knowledge movement, which directly draws from free/open source culture and tradition. In the late twentieth century, many thought that open collaboration initiatives would forever alter nearly every industry and, possibly, change the very nature of capitalism.[4] Indeed, at that time, not only was Linux already a major open source player dominating the markets of server and smartphone operating systems, but Wikipedia was also easily winning the competition with Britannica and has since undoubtedly conquered the online encyclopedia niche.[5]

Now, however, we clearly see that these successes were isolated, and the utopian idea that all industries would soon follow suit was likely but a dream. Even though there are plenty of promising, growing, and expanding grassroots initiatives and an open collaboration model in the brick-and-mortar world of manufacturing, none are yet as successful as their commercial counterparts.[6] Moreover, corporations have bit-by-bit taken back some of the successes of the free/open source movement. For instance, Linux dominates the smartphone market but

Open collaboration is a form of organization and cooperation in which participants share a common goal but are loosely coordinated, yet together they create a product or service and make the final result available to anyone interested.

does so in its Android variation, which Google developed in such a way as to promote its own services. Because simple open-source Android comes with few apps and functions expected by contemporary users, such as Maps or a modern email client, Google piggybacked on Android's success; it effectively forced nearly all Android phone producers to pay for a Google-branded Android bundle license, which brings Google Maps, Gmail, and Play Store to the table. In other words, Google exploited the system and reversed its openness, thus reducing it to a freemium platform.[7]

Similarly, even though Wikipedia does not have much direct competition, Google reuses vast amounts of the online encyclopedia's content in the right sidebar of Google searches (also known as the Knowledge Graph panel). In many cases of commercial reuse of Wikipedia, especially in the increasingly popular voice interfaces, users rarely know that the answers to their questions come from Wikipedia. Such use, which may deter readers from accessing the site if the Google sidebar supplied sufficient information, significantly limits not only Wikipedia's ability to raise funds, but also to attract new users. Introducing new users is essential to the site for several reasons: because only a typically small fraction of Wikipedia readers volunteer to write and edit its entries, and there is a natural burnout; and because volunteer engagement in crowd-sourced movements escalates slowly, through gradually increased participation.[8] Overall, commercial enterprises adapted to the free/open source and open collaboration environment by taking advantage of them rather than giving much back, or by sharing whatever benefits they received as a result of using the openly developed goods. Two recent acquisitions, first of GitHub by Microsoft and second of Red Hat by IBM—the latter for a mind-blowing sum of $34 billion—also show that traditional corporations not only adjust to peer production as a source of free labor but also appreciate platforms that allow for free software development and collaboration.

The Abuse of Sharing

The term used for new phenomena associated with open collaboration, *sharing economy*, sometimes referred to *collaborative economy*, has become a buzzword today, possibly the most recognizable term related to the processes of open collaboration. It has yielded about seven million results in a Google search as of May 2019, thanks to the seemingly ubiquitous coverage in mainstream media about the commercial success of Uber, Airbnb, and similar platforms, and also as a result of the term's prevalent use in academia.[9] Quite clearly, however, the name is doubly misleading. As we show in this chapter, what many often call "sharing economy" is neither about sharing nor just about the economy. In this section we offer a brief typology of commonly used terms related to the topic and categorize them based on the monetary or nonmonetary orientation of platforms and users.

Let's begin by looking into the premise behind "sharing." As we commonly understand it, sharing involves no monetary exchange: this is certainly true of Linux or OpenStreetMap but clearly not of Uber or Airbnb. Here's an example of sharing from everyday life to further illustrate our point: If two roommates use the same hairdryer, they share. If one of them wants to charge the other for the privilege, we can safely predict that this refusal to share is likely to change the dynamic between the roommates. On this basis we can argue that corporations such as Airbnb or Uber undermine the preexisting sharing culture in many aspects. Think about how the bottled water industry put a price tag on what was previously perceived as a common good.[10]

Airbnb and Uber, for their part, have appropriated the vocabulary of sharing and collaboration to piggyback on the popular notions of those terms. It even makes some sense, as early definitions assumed the backbone of the sharing economy to derive from leveraging surplus and talent goods.[11,12] But let's look at what happens when Uber markets a service—and assigns a precise monetary value to it—by making it seem analogous to giving a friend a lift. This effectively transforms the social context of what used

The term used for new phenomena associated with open collaboration, *sharing economy*, sometimes referred to as *collaborative economy*, has become a buzzword today.

to be a favor and turns it into something to be bought or sold. In this sense we can more adequately portray Airbnb and Uber as contributors to an *unsharing economy* because what they offer relies on blatant commodification. It is difficult to disguise how little their services have in common with actual sharing. They are in fact an outcome of a typical Silicon Valley innovation narrative.[13]

Nicholas John argues that pointing to the misuse of the word *sharing* glosses over a more important issue: what we can learn from the myriad ways we actually do use the word. For instance, business people can say the phrase "sharing for money" and understand it perfectly well.[14] And yet seeing the common-sense contradiction in such a use exposes corporate newspeak. Even though the critique of "actual sharing" may be a straw man argument, as no profit-oriented economy is really oriented toward sharing, it is nevertheless worthwhile to observe that the business discourse and appropriation of the term are largely marketing ploys.[15]

Philippe Aigrain argues that the new era of sharing, especially with regard to digital files, has a transformative power to alter the society and its markets.[16] In some cases, sharing could also rely on the cooperativist principle and facilitate joint ownership of the platform or create common rules for its governance. But neither scenario holds true for Airbnb or Uber: they offer a fixed-rules platform that relies on creating an entirely nonflexible, uniform, and closed model. Uber even wants to exert tight control over laborers and users rather than provide ways to empower them. In China, Uber monitors its workforce to see who participates in protests, whereas in countries everywhere it uses algorithms to determine if drivers work for the competition.[17] At the same time, companies who host sharing platforms desperately avoid state regulations by insisting that they are technology (not service) providers—intermediating platforms, they say, rather than running an accommodation rental service (in the case of Airbnb) or a taxi company (Uber). They champion deregulation, which shifts the liabilities and responsibilities of business to workers, depriving them of whatever job security they could formerly count on.[18] As a result, these platforms can still avoid responsibility for bookings and cancellations, or for enforcing antidiscriminatory regulations. Public and industry pushback, however, may eventually inhibit such loopholes.

Let's momentarily ignore this conundrum and stretch our understanding of the sharing economy to include monetary transactions. And let's agree as well that Airbnb (for example) fosters apartment sharing among strangers. The extent of this endeavor as a whole in no way supplants the hotel industry or private apartment rental. But it does introduce new elements, including access to platforms that not only connect interested newcomers or potential users but also provide tools to foster trust through ratings and evaluations. Initially these platforms seemed promising as a means to boost the potential of underutilized physical assets and address other problems related to "idle capacity."[19] But the economic climate has changed in the decade since the end of the financial crisis in 2008. The numbers of those who buy rental properties and cars specifically to participate in the new "sharing" platforms are currently on the rise, and these platforms are essentially creating a new business model. As Nick Srnicek notes, "Phenomena that appear to be radical novelties may, in historical light, reveal themselves to be simple continuities." Srnicek goes on to explain that despite capitalism's historic, incredible flexibility, its "invariant features" and the "imperatives and constraints it imposes upon enterprises and workers" must be seen in light of changing social relationships, particularly regarding property and the market.[20] Coming back to the present day, these platforms admittedly resolve the problem of coordination costs, allowing untapped resources to enter the market. Nevertheless, these resources have already been paid for and removed from the market; hence, the platforms allow for their endless recirculation.

An Umbrella Term of Contradictions

Should we agree that sharing economy is a meaningful term and covers initiatives as different as Wikipedia and

Airbnb, we would immediately be pulled into a jumble of contradiction. We would have to recognize sharing economy as something in between "a pathway to sustainability or a nightmarish form of neoliberal capitalism."[21] Indeed, sharing economy "constitutes an apparent paradox, framed as both part of the capitalist economy and as an alternative."[22] That is, sharing economy has become an umbrella term for many (sometimes conflicting) ideas.[23] Although the term could prove useful as we reflect upon various aspects of the social turn toward collective activity and away from mediating institutions, the overgeneralized or fuzzy conceptual chaos surrounding it impedes that opportunity. As Arun Sundarajan observes, "sharing economy spans the market-to-gift spectrum."[24] It includes cooperativism, gift giving, barters, and other altruistic ventures, as well as strictly commercial, for-profit endeavors that would make Henry Ford look like a socialist. Sundarajan's approach assumes "sharing economy" to be a catchall concept, in which the logic of market exchange and functional efficiency somehow—in a truly Hegelian fashion—coexist with the logic of communal virtue and symbolic socialization.[25] Like the two-realities paradox in the Schrödinger's cat thought-experiment, sharing economy simultaneously saves us from hyper-consumerist obsessions of late capitalism while serving steroids to neoliberalism.[26]

It is more reasonable to observe the clear distinction between platforms: first, those that adapt new technologies and harness them into the traditional capitalist system; and second, those that use technology to actively undermine this system. The first platform type leverages the capabilities of modern technology to return to nineteenth-century industry standards in labor relations. In effect this creates a post-Taylorist ideal: the increasingly strict control over the means of production and a sharp decline in workers' rights. Applying this to the case of ride platforms like Uber, when municipal regulations in the taxi industry began to unravel and control radically shifted to the companies, drivers suffered ever-deeper levels of precarity.[27]

The CEO of TaskRabbit once openly declared that the company is hoping to "revolutionize the world's labor force."[28] Indeed, in a truly revolutionary fashion, it appears that this change partly relies on obliterating the achievements of unions thus far in their struggle to secure basic mutual obligations in worker-employer relations. For instance, by insisting that Uber drivers are "contractors" rather than "employees," Uber can avoid minimum-wage requirements, health insurance, and retirement plans for its drivers, even though it can exert control and impose requirements exceeding those typical in a regular taxi corporation. This process characterizes what is often (and more accurately) called a *gig economy*: an entirely contingent workforce, available on-demand, and—thanks to advanced IT systems—ready to fulfill the "just-in-time" philosophy of workflow and inventory.[29] (See figure 1 for a comparison of interest shown in the terms *gig economy* and *sharing economy*.) A contingent workforce, by the way, operates with radical information asymmetry: in this context specifically, workers are at a disadvantage because they deal with corporations that create an extensive system of gamification and psychological trickery to coerce "contractors" into working longer hours.[30,31] A contingent workforce also faces an unprecedented power asymmetry that allows corporations to intercept most of the value generated by independent workers.[32] Moreover, that exploitation is highly gendered and racialized.[33] Silicon Valley is using technology to build a new global underclass, forced to perform what Mary L. Gray and Siddharth Suri aptly call "ghost work," dehumanized and invisible labor streamlined by apps and intermediary platforms.[34]

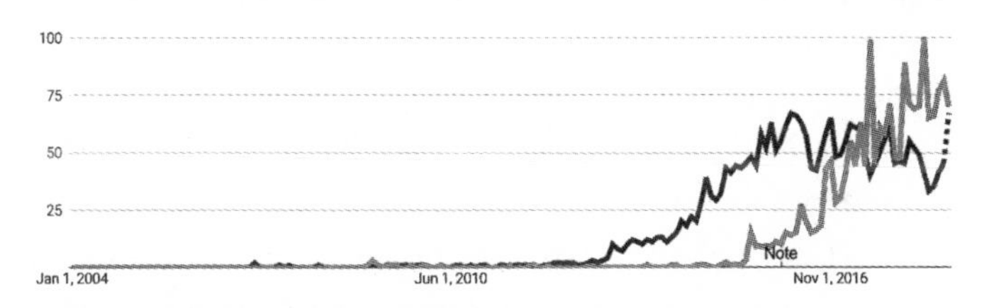

Figure 1 A Google Trends graph measures interest in the sharing economy (in black) versus the gig economy (in gray) according to search engine statistics from 2014–2019. Source: https://g.co/trends/TvMAB

Because of these practices and conditions, some authors propose the use of *platform capitalism* to refer to the ongoing phenomena of corporate domination in the "sharing economy."[35] This term well reflects the economic drive of the capital market to harness the power of cutting-edge technologies, disempower users, and turn them into workers and customers, rather than inspire them to actually share anything. Of course, it is possible to perceive the domination of the sharing economy by commercially driven organizations as a temporary distortion of the idea.[36] It may well be that corporations are the first to adapt to new technologies and platforms because they are driven by profit, but eventually nonprofit initiatives may outpace them. And yet this seems unlikely. Profit-driven organizations exert great power in a large number of industries; they test different models and mechanisms to gain a competitive edge over sharing economy "incumbents," including "tight or loose control over participants and high or low rivalry between the participants."[37] The motivation of the participants themselves is another important factor to consider: as we previously explained, if organizations focus on profit or even profit maximization, they do not share.

At the same time, online cooperativism provides viable alternatives to many contemporary participants and organizations consciously trying to avoid the exacerbating effects of platform capitalism.[38] Their actions starkly contrast with the pseudo-sharing approach typical of the second platform type we identified earlier, in which technology is used to actively undermine a capitalistic system that relies on profit motives, the expectation of reciprocity, and the lack of communal feeling.[39] The Green Taxi Initiative in Denver, a worker-owned alternative to Uber, is one such example with a focus on cooperative sharing. Some call these ventures *platform cooperativism* because they depend on digital technologies to intermediate between dispersed clients and a network of service providers while emphasizing collaboration and fairness, not only profit generation.[40,41] Platform cooperatives rely on shared responsibility for governance and more equitable distribution of profits because involvement in the organization and its financial outcome distinguishes them from platform capitalism.

Some free/open source initiatives popularly considered part of a "sharing economy" organize around sharing communities. They often rely on commons-based *peer production*, which includes open-source endeavors like Wikipedia or Linux. Peer production organizations do not pay their contributors, who participate to learn and earn a reputation—whether professional or communal—as well as to do something good for society.[42] Because of this orientation, many peer production enterprises are nonprofits, a form of organization that better suits the contributors' philosophy. More recently, especially when it comes to for-profit peer production, organizations increasingly rely on gamification, as seen, for example, in the Google Guides community (see chapter 3).

Producing Shared Value

We devote all of chapter 3 to a discussion of peer production's important role in the transformation of collaborative society, and especially to how it embodies aspects of the ongoing revolution. But let's look now at how the focus on "production" emphasizes the creation of goods or services that possess either some concrete value (as with Wikipedia) or potential monetary value (for instance, Yelp). Yet such a focus overlooks other endeavors, like Couchsurfing or Airbnb, typically associated with the ongoing, major social turn toward collaboration, and it omits as well a consideration of the performative aspects of this change. Production may be a side effect of content-sharing portals such as 9GAG or Imgur, even if production is the main reason for their popularity. These new initiatives grow out of the fun we derive from cooperating with others, sometimes with no concrete end product at all. Take the game *Minecraft* or the programming language Scratch, for instance, where the joys of performing a task together may well prevail over the concrete outcomes of individual projects.[43] Some authors who recognize the importance of the cooperative process propose the term *collaborative consumption*.[44] But we believe that consumption, like production, is not best word with which to capture the value of changes we're experiencing as a result of the so-called sharing economy.

Indeed, the second part of the problem with labeling these new and complex social changes as "sharing economy" comes with the focus on the economy and market terminology itself. A lot of the visible shifts occurring because of Wikipedia, Linux, or TripAdvisor, for example, far exceed what this vocabulary can convey. Admittedly, much of the disruption occurs on the business side of these operations, the economic ramifications are surely huge, and examples abound. For instance, New York City taxi medallions required for operating and owning a yellow cab in the city cost more than a million dollars each at their peak price in 2014, and even then were considered a good investment. Now Uber and Lyft have upended the market for (and the means of) "hailing" taxis. As a result, medallions are an unbearable financial burden to many owners who cannot compete with apps that impose fewer rules governing fares and vehicle equipment.[45] Similarly, the rapid growth of Airbnb has wreaked havoc in the hotel industry and the apartment rental market. TaskRabbit, for its part, made the process of hiring a plumber similar to visiting an online dating site. Its reviews and ratings feature may have removed the hit-or-miss factor in finding competent handyman services. But the company's recent policy changes on repeat-customers discounts, and an increase in the fees Taskers themselves must pay to the company, have not much lessened the workers' job precarity.[46]

Still, there also emerge many phenomena clearly related to sharing and cooperativism, but not necessarily best explained as economic. They rely on the human drive to exchange knowledge, discuss, and help even though such action might not always produce any concrete output. The rise of peer-to-peer (p2p) lending platforms, for instance, has been possible thanks to the internet. Kiva is one such nonprofit platform that offers microloans to people from underserved groups all over the world; minimal interest paid by the borrowers goes only toward defraying operating expenses of intermediary field partners.

Other examples follow, many of which we examine in detail in later chapters: The citizen science movement opens up the potential for nonprofessional researchers to advance findings and discoveries in numerous diverse fields; in public health, for instance, patients may partner with doctors to identify community health concerns or to implement solutions.[47] Determined amateur enthusiasts who expect no financial compensation develop new methods and processes of detecting and reducing smog or radiation.[48] Cooperative communities of biohackers strike a balance between activism and entrepreneurship by performing experiments on themselves.[49] The Quantified Self trend encourages people to measure their various biological metrics—in the simplest form including heart rate, the number of steps taken per day, or calories burned—and potentially share the data with members of the community to motivate and change the behaviors of others.

Moving away from the realm of science and medicine, millions of users participate in communities that practice torrent file sharing by uploading already downloaded files so others can download them (a process known as "seeding"), and by doing so they risk prosecution for allowing p2p-pirated media distribution.[50] Couchsurfing, although operated by a for-profit company, serves a community of nonprofit participants willing to host strangers from all over the world. Large social movements, such as #MeToo, Occupy Wall Street, or the Arab Spring, organized through social network platforms and influenced major social revolutions. People from all over the world engage in modifying, exchanging, and republishing remixes of pictures, movie excerpts, or works of art as part of a crosscultural collective.[51] Many social networking platforms, even the broad-based Facebook, allow users to give away things they do not need anymore to strangers, and their popularity shows the extent to which people are willing to act for others without direct gratification. The nonreciprocal exchanges, social bonding, and solidarity are powerful drivers of the new, collaborative turn.[52]

A quantitative analysis of the academic literature about the sharing economy shows an equal focus on four main areas: sustainable development and commons governance; community building and participation; labor, production, and digital markets; and new production and consumption models.[53]

A clear divide exists not only between capitalist and cooperativist platforms but also between users oriented toward profit maximization or nonprofit status. The commonly used terms shown below in bold often align themselves according to this divide:

Table 2.1

	Profit-maximization-oriented users/producers	**Nonprofit-oriented users/producers**
Capitalist platform	**Platform capitalism, gig economy** Uber, Airbnb, TaskRabbit, Turo	**Collaborative economy** TripAdvisor, Yelp, Couchsurfing, 9GAG **Quantified Self**
Cooperativist platform	**Online cooperatives** Time banks (Timebank.cc, Hourworld.org) Food co-ops Peer-to-peer lending	**Peer production** Wikipedia, Linux, OpenStreetMap **Citizen science** **Online activism** #MeToo, Occupy Wall Street **Biohacking**

We believe that many of these examples refer to the phenomena of open collaboration or collaborative consumption, but neither term, sharing economy nor peer production, adequately covers their scope. As we see it, the significant change occurring in the society, apparent in the above phenomena, relates first and foremost to people's universal drive to collaborate; we also believe that this drive has been significantly magnified and enabled by the new internet platforms and tools. Collaboration may sometimes involve sharing and sometimes rely on doing things together or discussing. We moreover believe that this change transcends the economy and enters many other aspects of life. Although technology-enabled sharing and collaboration affects modes of scaled distribution, it vitally relies on social intensification and predominantly serves the purposes of connectivity and collective formation.[54] We thus propose that the social turn underway is related to the rise of collaborative society.

The Possibilities of Giving

Sharing economy, open collaboration, peer production—all these terms describe aspects of collaborative society that have scaled up globally thanks to technology. The turn to collaborative society is significant because it not only redefines some aspects of our economic system but also challenges the knowledge hierarchy, the system of professions, and even the very understanding of authorship and ownership of media. The dominant logic of late capitalism relies on mediating as many facets of human relationships as possible and turning them into services. Platform capitalism naturally extends this set of values, as does the perception that most relations are economic by nature and defined by their possible monetary value. What we observe now, though, is a turn toward peer-to-peer relationships and exchanges that circumvent this logic. As Eric S. Raymond notes, the dominant capitalist exchange economy relies on adapting to scarcity, whereas gift culture relies on adapting to abundance.[55] Raymond perceives early free/open source development and hacking as clearly about the latter. Sharing is an alternative form of redistributing goods and services.[56] Through a large set of technologies, collaborative society builds on this alternative, allowing more and more departures from the traditional capitalist model.

Platform capitalism may soon see its demise due to blockchain or other peer-to-peer technology development, especially as p2p empowers users with means of direct interaction that eliminate the need for intermediary platforms. Collaborative society, however, is a phenomenon with much deeper roots and possibly wider ramifications.

Nevertheless, collaborative society faces major challenges. On the one hand, it experiences internal tensions between capitalist and nonprofit tendencies—whether they attempt to use new technologies and social systems to advance exploitative capitalism, or to undermine the capitalist system as a whole. On the other hand, the possibilities to cooperate that emerge from collaborative society simplify the potential for helping others, either on a pro bono basis or for profit. In some industries, collaborative society endeavors involving payment or financial cooperation have not been popular. For instance, Knol, a Google-run encyclopedia that planned to pay contributors like YouTube never gained traction and failed to launch. In other industries, for-profit contributions are much more typical. There also remain a large number of initiatives run by nonprofit-oriented people through cooperativist platforms. Some foster cooperation, others encourage individual work. Some are run by users, others serve them. Some promote cooperativism and altruism whereas others enable market exchanges. But as Kieran Healey observes, "Just as a market relationship does not automatically make for exploitation, neither does a gift relationship automatically create solidarity."[57]

In this book, we want to showcase these phenomena. We want to put a particular emphasis on collaborative society beyond the much-touted platform capitalism. And we focus on cooperative behaviors—enabled as they are by new technologies—that occur without a dominant financial motive. Because they receive much less attention than they deserve, our goal is to give them their due as the primal force behind many of the changes in modern society.

3

PEER PRODUCTION

According to Wikipedia in 2019, "*peer production* (also known as *mass collaboration*) is a way of producing goods and services that relies on self-organizing communities of individuals. In such communities, the labor of a large number of people is coordinated towards a shared outcome."[1] Wikipedia itself is one of the most classic examples of a peer production project. It also is commons-based, meaning that it results in the production of common goods—unlike, for example, Facebook. To some extent, Wikipedia also relies on peer production of content and social relations, but not with a stable, permanent, and publicly shared outcome. In our view, however, peer production particularly fits those collective efforts that aim at solving a problem common to those who participate.[2] We see the value of peer production as a means to target serious and

According to Wikipedia in 2019, *peer production* (also known as *mass collaboration*) is a way of producing goods and services that relies on self-organizing communities of individuals.

more long-term solutions—in contrast to online leisure and creative collaborative projects.

Some scholars prefer to view peer production as a phenomenon more typical to online cooperatives that produce free and open outcomes, and thus often use the term *commons-based peer production* as a way to contrast it with profit-driven peer production.[3] In this understanding, it may be a vehicle for redefining capitalism and introducing new models of social value.[4] Arguably, peer production appears as the most significant organizational innovation resulting from online cooperative technologies.[5] In more modern applications, the peer production approach can be applied to physical goods manufacturing in a "design global, manufacture local" model, which also seeks to challenge the existing political economy and further sustainability through degrowth.[6]

Furthermore, the phenomenon of peer production can be understood as a technically enabled way to run a social movement focused on creating something for the common good.[7] And in that case it can be strongly motivated by ideology.[8] An important part of many commons-based peer production initiatives involves a social turn toward more direct interaction and away from mediating institutions. Sociality and collaborativeness inherently drive peer production and often counteract the logic of capitalist models that rely on singling out consumers.

This sociality very rarely comes with anonymity. Peer production contributors usually identify themselves by a user name and have credibility in their community. According to Arun Sundarajan, premodern trade depended on community-driven reputation, and the industrial era relied on the trustworthiness of institutions that mediated transactions. Currently, however, a creative hybrid of former trust-enhancing mechanisms has taken over: the user-generated reviews of fellow peers, along with brand reputation, allow us to make better judgments when conducting transactions.[9]

We believe that this trust shift has indeed occurred, but observe that it differs from institutional trust: peer production organizations substitute the trust in individuals with the trust in procedures and the process as a whole. This trend is most visible in blockchain technology, but it's clear as well in regular website platforms. As a result, individual credentials may not matter. For instance, a medical doctor with 20 years of experience may still need to support arguments and provide pertinent research when persuading those who contradict a particular diagnosis. In the case of Wikipedia, this makes a lot of sense: as long as information added to the site has sufficient proper references and relies on credible sources, a neutral viewpoint, and grammatically agreeable language, it doesn't matter what academic degree its editor holds.

Consensual Authority and Potential Conflict

Authority and social standing definitely do exist in the realm of peer production, but such status derives from contributing to the project. For instance, Linus Torvalds has the final authority when it comes to incorporating modifications and additions to the Linux kernel. Although Torvalds prefers to keep a low public profile and a neutral stance regarding competing software products, he is a strong leader who has been known to speak bluntly when confronted by views opposing Linux or open-software practices. He not only created the operating system that started the project in 1991 but has since been one of its leading developers; by 2006 he was responsible for as much as 2 percent of the whole code (a substantial amount, given the complexity of a code to which thousands of other programmers have contributed).[10] Torvalds basically validated his leadership through long-term commitment and contributions. In contrast, Jimmy Wales also had tremendous authority on the English Wikipedia as its founder and major early contributor. Yet when Wales expressed his views about Wikimedia projects that he did not organize, such as Commons and Wikiversity, he faced strong opposition and was even forced to give up some of his technical privileges.[11] Part of the opposition may have come from his perceived lack of participation, because in the Wikimedia community "one's edit count is a sort of coin of the realm," and one's status in one project does not transfer to any other.[12] Moreover, status does not last forever. A prolific past contributor who no longer participates actively will likely have little informal authority. Currently, Wales serves on the Wikimedia Foundation Board of Trustees and seldom engages in the daily decisions of particular communities. Notably, many peer production communities establish supporting organizations, often in the form of foundations, to address their most basic needs and general development; these organizations can adopt more traditional governance models.[13]

Apart from its unusual approach to authority and credentials, peer production often perplexes organization scholars because of its apparent lack of formal hierarchies and its minimal governance and leadership, all of which contrast with the indisputably effective results of collective work.[14] In fact, many peer production organizations have a strong anti-hierarchical mindset, and a distributed leadership or even a leaderless ethos.[15] These communities sometimes face a paradox then: while relying on a heterarchial, structureless, and leaderless approach themselves, they outsource daily operations and business procedures to entities with clear hierarchies, structures, and a pecking order.

The lack of hierarchies within communities can yield various positive side effects. For instance, newcomers have the same formal status as veterans. Newcomers

Apart from its unusual approach to authority and credentials, peer production perplexes organization scholars because of its apparent lack of formal hierarchies and its minimal governance and leadership, which contrast with the indisputably effective results of collective work.

might lack the social capital that comes from peer recognition, but their ideas will not be discarded on the grounds of short tenure. Moreover, even if veterans have functions of power, such as an administrative role on Wikipedia, they are expected to treat newcomers with the same care, respect, and attention as everyone else. Furthermore, no one may dictate what others should do. Although users must follow the policies, no one expects them to work on a specific task. Even if everyone agrees that the project definitely needs a modification or addition, and there are users who could easily provide it, others may only ask them politely to consider stepping in, and will not interpret their refusal negatively. As a result, topics and tasks in peer production are usually covered unevenly because the individual efforts that address them or carry them out are conducted with little overall guidance and governance. Yet participants will gladly help each other when requested.[16]

Peer production relies on the principle of open collaboration: everyone is free to join and help, as well as leave at any point. This option is important because it strongly influences the social organization of work. First, it means that only very disruptive participants will be banned from contributing, and even when banned, they can still return, either under different user names or anonymously. In one case, administrators indefinitely banned the most prolific editor on Polish Wikipedia for his hostility to newcomers, but he still regularly contributes to Wikipedia and tags his edits in a way that makes it possible to recognize his authorship. Even though there were discussions among Polish Wikipedia administrators about reversing all his edits as a matter of principle, pragmatism prevailed and they agreed that constructive contributions should remain. Second, because users may leave when they please or just refrain from contributing, it's difficult to rely on a more structured division of labor. It's impossible to presume, for example, that a given administrator will be available to deal with a problem. Consequently, responsibility shifts to structures and roles. Although the community cannot expect a particular individual to fulfill a duty, it may rely on any functionary to perform the required task when needed.

Without a clear hierarchy, the community often makes decisions via a participatory process. Hence, it highly values consensus. Because everyone is a volunteer, decision-making tools that privilege the majority won't work because keeping the minority involved, motivated—and (ideally) persuaded to adhere to the majority view—is a top priority. This is why many peer production communities take steps to ensure that participants adopt a given view and at the same time include all interested parties. In highly collaboration-oriented communities there may be less insistence on consensus building.[17] But the general focus is clearly on agreement.

For this reason, collaborative communities strongly encourage participants to express their opinions. In fact, based on one of the principles of consensus-based decision-making, if no one objects, an agreement has been reached. When combined with the heterarchical organization and a democratic ethos, this process may lead to lengthy discussions on the most trivial subjects, and they may be intense and harsh in tone. In fact, basic politeness may not necessarily be considered a virtue. In many projects, a strong ethos exists to support putting aside ego, skipping pleasantries, and bluntly discussing the issues. Linus Torvalds expressed this sentiment quite clearly when he said he likes to argue and is "not a huge believer in politeness and sensitivity being preferable over bluntly letting people know your feelings." To further make his point he added, "I'm not a nice person and I don't care about you."[18]

To people who are not used to the level and style of discussion typical in peer production projects, it may appear that everyone engages in permanent, all-out conflict with everyone else, even though what they observe comes partly from the lack of hierarchy, and even though, in many cases, explicit norms exist that forbid open hostility and profanity. For instance, Wikipedia has a policy requiring civil behavior and forbidding personal attacks; administrators temporarily block violators. Nevertheless, the geeky focus on IT tasks, the casual disregard for basic empathy, the dispassionate ethos, and the depersonalization and dehumanization of interactions often lead people to think of the community as an unfriendly environment, especially since interactions are conducted in writing and mostly anonymously.[19] In a way, the effects of these conflict-driven factors cement the hermetic editor environment and help the model work, even though they prevent its further development. Unsurprisingly, a conflict-driven environment and a clear gender bias among many editors result in a huge gender gap.[20] Although statistics vary, it is safe to estimate that less than 20 percent of English Wikipedia editors are female.[21] Nonetheless, gender proportions among readers are evenly split. Such an imbalance even more radically surfaces in free/open source software projects, where less than 2 percent of developers are female.[22]

Conflicts in peer production projects are also typical because they are addictive; indeed some take the hyperbolic view that conflict at Wikipedia is "as addictive as cocaine."[23] The need to win in an online discussion, using Wikipedia as an example, derives from natural instincts to compete. In peer production communities this drive is channeled into something productive, because the only way to really prevail is to add to and develop the project. Should both sides of the conflict be determined, the commitment quickly escalates. Sometimes the conflict topics may seem absurd to bystanders. For instance, a long debate on Wikipedia—counted in several years and hundreds of thousands of words—focused on whether the portal for the well-known Polish city should use the name Gdańsk or Danzig. Many other trivial issues prompted lengthy discussions, such as the spelling of the title in the article about "yoghurt" (or "yogurt"), or whether Mexico has an official language.[24] The conflict-driven character of such interactions may help keep people engaged but can produce negative side effects. Any decision, even the most inconsequential, will require much "back and forth" on the issue, which makes the introduction of major reforms difficult. Even the most obviously right decision will surely receive criticism. Moreover, the whole setting affects the profile of users who stay in the community: they are the ones who wallow in disputes. Since conflicts are typical and the work of editing an encyclopedia or writing code is still hard work—even though much fun—burnout among peer production project members abounds.[25]

Assessing Freedoms in Creative Cooperatives

Many of the characteristics we've described for peer production initiatives also pertain to free/open source initiatives. In fact, in the beginning of open collaboration movements, many of them closely copied the principles and culture typical to free/open source software. These principles assume that the outcome of peer production should be free in the sense of not requiring any payment for use, but also free in the sense described in 1986 by Richard Stallman, when he explained the philosophy of the Free Software Foundation: free to use, free to study and change, and free to redistribute with or without changes.[26] It is free in the sense of "free speech" as well as in the sense of "free beer."[27] These freedoms enable forking, which we can think of in visual, nontechnical terms as the way a river can go off (or change) course.[28] This right to fork is important, as it actually helps the whole ecosystem: Let's say we want to do things differently and better; we may try to persuade others to follow us. If we succeed, what surfaces may often be something better indeed. If we fail—well, the original project is still there to return to.

The freedom to redistribute, however, legally allows for a collaboratively developed product to be sold or repurposed. As a side effect, this ability has resulted in an entirely new business model of e-books produced "by repackaging public domain content scraped from the web": on Amazon, there are thousands of books that are basically Project Gutenberg rip-offs or Wikipedia articles on related subjects gathered in one file.[29]

The ideological commitment to these freedoms can be very strong. For instance, although a 2002 idea that Wikipedia run ads only surfaced as a suggestion, a major outcry in the community ensued, and the majority of

Spanish Wikipedians forked their own online encyclopedia called Libre.[30] As a result of the tension, the community established the Wikimedia Foundation, which irrevocably established Wikipedia as a nonprofit endeavor. Similarly, when Oracle overtook Sun and OpenOffice in 2010, its users forked into LibreOffice because of their concerns regarding licensing and its potential limitations. When Wikitravel introduced ads in 2012, a large part of its community forked into Wikivoyage, a new project supported by the Wikimedia Foundation.

Currently, however, many for-profit corporations successfully introduce peer production—not commons-based production—as a method of generating collaborative outcomes, while keeping the content non-open and clearly ignoring the free/open source software principles. The examples of Yelp, Goodreads, TripAdvisor, or Google Guides communities show that the free and open ideals, important for many communities in the past, are no longer the deciding factors for contributors when in determining the attractiveness of online projects. The reasons for denouncing non-open or non-free collaborative formats seem related instead to the projects' value integrity: communities that believed their work would be free and open objected to any possible changes, while communities that participated in commercial and closed platforms were fine with the potential for change from the beginning. Moreover, the whole climate of support for free/

open source may have possibly changed with the growing popularity of freemium and other business models.[31] For-profit organizations may have finally grasped how to organize communities for peer production with the right incentives, just as they have found ways to influence many independent free/open source software projects.[32] But it seems that freedoms have never been truly decisive determinants of massive collaboration. Interestingly though, the peer production approach proved effective against the traditional model: for instance, TripAdvisor is more popular than Lonely Planet or Fodor's, while Yelp overshadows the classical Yellow Pages.[33]

Nevertheless, the paramount peer production successes like Wikipedia, Linux, Project Gutenberg, or Mozilla also happen to be free to copy, study, reuse, or change. Let's take a closer look at such initiatives. They can help us better understand the new collaborative society, and also inspire many less radical structural designs and solutions in more traditional organizations. These projects share many characteristics of organizational culture as well, because they are all inspired by hacker culture and have roots in Unix.[34]

It should not come as a surprise then that peer production challenges conventional economic theories of motivation because its lacks clear extrinsic incentives. In projects that require professional skills such as programming—an indispensable skill for free/open source software development—active participation may partly stem from the need to build a portfolio, develop a reputation, or participate in networking exercises that would be useful in some future career.[35] And yet this does not apply to some peer production initiatives, such as Wikipedia. After all, "writing Wikipedia articles" is not a credential that many people can include on their résumé, and the current market for experienced encyclopedia writers is quite small. What counts, though, is the ability to participate in a community of people with similar ideals,[36] even if the very understanding of the word "community" differs between users.[37] This drive also appears in free/open source software projects: the need to collaborate with other similar-minded people is arguably more important than just working on individual, professional credibility. Nevertheless, when building a professional reputation the engagement in collaborative tasks and communal experience is quite crucial,[38] whereas the focus on individual career-orientation correlates negatively with participation and longevity in peer production projects.[39] What matters more is the collective activity.

Risks of Outsourcing from the Playground

The experiences of community and collective activity undoubtedly bring much joy, and with it the potential for such activity to become addictive. Although participation in peer production is entirely voluntary, important questions arise about control and exploitation.[40] Some users, for instance, satisfy their social needs by engaging more and more in peer production, which is great for the project's development, but potentially also dangerous for the individuals involved. The internet then becomes both "playground and factory" and a large resource of free labor.[41,42] Considering that the results of this unpaid, voluntary work are material, we must ask whether the benefits enter the common pool or build a commercial competitive advantage when producers and consumers become one and the same, or *prosumers*. As George Ritzer and Nathan Jurgenson warn, prosumer capitalism, in its dependence on digital technologies, relies on new forms of exploitation through unpaid work gamified as fun.[43]

A study of the OpenStreetMap project—run by a community of enthusiasts who collectively create maps using Apache software on an open license—shows that participation in (and as) a community rank high in the motivations of active users.[44] But in many cases the "participation" does not necessarily mean doing something simultaneously together, with a division of labor and an awareness of other participants. Rather, as a study of nearly all Apache projects shows, peer production contributions are typically solitary endeavors, and input is often noncollaborative.[45] Most of the collaboration occurs when establishing common standards of work and the rules of governance. It may sound paradoxical, but open collaboration in the case of peer production relies much more on discussing how things should be done than actually doing them together. This may be so—in keeping with Elinor Ostrom's principles of stable common-pool resource management—because it requires "collective-choice arrangements that allow most resource appropriators to participate in the decision-making process."[46] Creating rules gives the participants a feeling of ownership and also fosters natural, informal hierarchies, important in the absence of formal ones, based on procedural proficiency.

One of the side effects of this phenomenon is that peer production communities may be exposed to bureaucratic dogma. For instance, English Wikipedia has over 50 official policies, containing nearly 150,000 words—about 3 times as many as this book—and more than 450 other regulatory guidelines and interpretive essays.[47] Some of this verbosity generates from new users who want to make the project their own by debating the rules and procedures. As a result, an arbitration committee serves as a collegiate court for dispute resolution.[48] It is quite natural as well that collaborative communities need to enforce certain standards of practice and may tend to impose constraints on the forms of participation.[49] Still, there are no simple mechanisms for eliminating the regulations or reducing in some other way the procedures that accrue. In the long term, this process will unfortunately lead to informational power asymmetry as people better versed in procedures gain the upper hand in the discussion, and will create as well an artificial entry threshold to join peer production projects.

Luckily, that threshold is very low. It is one of the leading reasons for the amazing success of this model. Peer production is based on compartmentalizing tasks into small, manageable tidbits. For example, writing a whole article or a whole software program can seem daunting, but when the task is to simply add a sentence or a line of code, the challenge is reduced and may even seem appealing. This approach enables community-based, collective knowledge creation organized as an iterative, ongoing process of gradual improvements.[50] The improvement part is crucial. Especially in the case of Wikipedia, more people are willing to improve a bad article than start a completely new and good one.[51] In other words, one of the motivations to contribute is because "someone is *wrong* on the internet."[52] All in all, though, peer production projects have one thing in common: a joint overarching goal that addresses a common problem. This is not so clear in the case of collaborative media consumption communities, which we discuss next, in chapter 4.

4

COLLABORATIVE MEDIA PRODUCTION AND CONSUMPTION

Apart from the semi-anonymous, collective creation of larger works and projects via peer production, a great deal of online collaboration occurs in the area of media, via short-term associations or impromptu cooperative stints, often aimed at immediate problems or issues. The methods of short-term collaborative media production are in many ways similar to those of long-term peer production, but they may differ in important ways.

The development of new communication technologies has made professional training—once a prerequisite—now irrelevant when it comes to creating, editing, publishing, and distributing media. Some describe this phenomenon as an outgrowth of *produsage*, a seamless blending of consumers and creators.[1] Indeed, we see a major change underway: the traditional model, in which active production takes a separate path from passive consumption, becomes less accepted (or effective) as the boundaries become less distinct. As Henry Jenkins argues, "Audiences, empowered by these new technologies, occupying a space at the intersection between old and new media, are demanding the right to participate within the culture."[2] Even though researchers don't always agree about the scale of the change, and the majority of users still do not actively create content,[3] the shift is noticeable even if only for its potential to overcome traditional divisions.

The increase in consumer production is possible not only because of technological changes—specifically, the creation of tools that allow collaborative work and publishing—but also because of social change. For instance, the rise of *copyleft* philosophy, which stems from the hacker and free/open source software subcultures, results in shifting boundaries around copyright, royalties, intellectual work, and attribution requirements; this of course makes reuse and derivative work much easier.[4,5,6,7] Similarly, while the new models of organizing work in ad hoc structures (and managing the work in a participative manner) rely on newly available technologies, their main impact and revolutionary effect lie in the social redefinition of organizational roles.[8,9]

The process of *collaborative media production*, sometimes called *commons-based peer production* (or consumer co-production), results in a radical redefinition of many professions and industries outside of knowledge or

The process of *collaborative media production*, sometimes called *commons-based peer production* (or consumer co-production), results in a radical redefinition of many professions and industries outside of knowledge or software production.

software production.[10,11] We see this when photographers lose ground to "amateurs" who distribute photos under open licenses for free or for a minimal stock charge.[12] Even though the demand for commercial photography still exists, the profit margins decrease because of massive oversupply, the availability of platforms that eliminate intermediaries, and the emergence of an informal media economy.[13] Similarly, journalists lose readers to bloggers and community-run portals.[14] And, as we discussed in chapter 3, encyclopedia editors and writers lose to Wikipedia.

The Cult of the Amateur, the Demise of the Pro

The decline in demand for a highly trained, skilled workforce is sometimes blamed for lowering the overall quality of outcomes in any number of fields. In media, Andrew Keen laments the demise of professionally generated content resulting from what he calls "the cult of the amateur," and he predicts a decline of our culture and economy resulting from the popularity of user-generated media.[15] In Keen's view, the oversupply of amateur content diminishes the financial viability for professionals to stay in business: nearly all content is free or almost free, and the costs of production no longer accommodate for the fact that the author has to make a living. In a way, produsage and technical revolutions have led to a massive deskilling of professional labor, to some extent resembling the Taylorist industrial revolution turn. The current ramifications of this deskilling, however, may have more than quantitative consequences: anonymous content creators cannot be held responsible for their work, as they do not have to follow professional standards. Their work may also be inferior, as the argument goes, but as a result of a process similar to how Gresham's law applies to money, "worse" commodities will nevertheless drive "better" ones out of the market.[16]

This scenario could potentially occur within platform capitalism organizations like Uber or Airbnb because their major market disruption may indeed rely on allowing amateurs to offer slightly more inferior services in competition with fully professional ones. But in the case of collaborative media production, the main issue is not necessarily related to quality. In occupations in which the main quality indicator depends on the customers' acceptance of the final outcome—in relation to aesthetic taste rather than intrinsic features—the buyers quite obviously have the agency to decide what level of quality they require at a given price. "Amateur" projects, however, sometimes provide good value for the money. Moreover, in many industries, the source of disruption stems from platform capitalism rather than the amateur character of work. For instance, Fotolia and Shutterstock serve as intermediaries for professional photographers to sell their photos, but they also radically drive the prices down for everyone, just as much as Spotify and Deezer do for professional music.[17]

Keen is right, though, in observing that prosumer culture may disrupt existing knowledge hierarchies. For instance, readers trusted Britannica because of its institutional authority. Wikipedia overturned this model by showing that knowledge authority may actually be vested in a transparent and solid process.[18] But Keen's critique assumes that before prosumerism, the division between professional and amateur content creators was sharp, which is an obvious fallacy. Talented amateurs have regularly transitioned into professional positions in photography, literature, journalism, and fine arts. The creative occupations have always attracted amateurs because of the playful character of the work and the focus on talent rather than formal training. Interestingly, many contemporary knowledge-intensive organizations, such as Google and 3M, incorporate playfulness as an accepted and standard method for making work both satisfying and attractive, even in quite mundane jobs.[19] With the "work as fun" paradigm continuing to emerge, the allure of such occupations may increase. And yet this paradigm, often associated with the gig economy, might negatively affect workers in an already precarious job market. The result is what some call "playbor," a neologism (play + labor) describing what happens when work and fun collide.[20]

The “work as fun” paradigm, often associated with the gig economy, might negatively affect workers in an already precarious job market.

The critique of prosumerism hits its mark when bloggers, who position themselves as independent but may in fact stand to benefit from corporate sponsorship, run a "flog" (a fake blog). Indeed, one of the problems with user-generated content is the lack of transparency and ethical standards that are well established in areas of publishing such as journalism. The rise of deepfakes (faked content, especially videos, produced by combining and superimposing images), as well as other easily accessible misinformation technologies, exacerbates this problem and makes ascertaining the trustworthiness of the source even more important. One response to this criticism is to create online collaboration communities that incorporate the help of professionals. For instance, Jimmy Wales, the founder of Wikipedia, created an independent for-profit company called WikiTribune, a community of volunteer editors who cooperate with hired journalists and provide the best of both worlds: fact-checking takes place under the auspices of collective news-gathering, and an authoritative, responsible, coherent writing style informs the content.

Of course the level of "amateur" quality may vary within or across industries. The best photos from Commons or Flickr might easily compete with "professional" platforms, but the number of amateur full-length movies remains very low, and as yet poses no threat to Hollywood. Perhaps even more fascinating, many internet users actually prefer "nonprofessional" work, sometimes as a result of its perceived authenticity. Online pornography relies heavily on amateur productions: portals that allow video sharing, such as Pornhub, RedTube, or YouPorn, experience a high demand for amateur videos. In 2017 alone, Pornhub reported that of the four million videos uploaded that year, 810,000 were amateur movies, and the "amateur" category is 27 percent more popular in Sweden than elsewhere in the world.[21]

Sometimes, amateur works gain popularity because they are actually better. This is often the case in knowledge and software peer development, as we discussed in chapter 3. For example, a study published in a 2005 issue of *Nature* revealed that Wikipedia and Britannica finished nearly neck and neck when it came to the number of errors in a sampling of articles from both sites concerning 42 different science topics.[22] Even though critics contested and discussed these results later, the "race" is long over now: Britannica went bankrupt, and Wikipedia has grown to be 85 times larger in word count than Britannica ever was. Current research of Wikipedia accuracy shows that it is on par or sometimes better than the "professional" resources, even in subjects as specialized as mental health.[23] By design, it is also much better referenced. Similarly, in its many variations, Linux is better than other operating systems for many commercial and noncommercial purposes.

The Joys of Collaborative Creation

Collaborative media creation may have a deeply therapeutic value when it operates as a forum for discussion. Take clientsfromhell.net as an example: the website presents a collection of funny narratives about experiences that designers have had with their most annoying or troublesome customers. Similar stories abound on sites geared for support specialists wanting to vent frustrations at customers who act unreasonably or just plain stupidly.[24]

Yet another reason for participating in collaborative media consumption is the pure fun of doing things together, even without any creative outcomes. The Pirate Bay, the largest torrenting website in the world, is a prime example. Saving money may be a factor for some who use the service for downloading, but the economic incentives to keep sharing the torrents are minimal compared to the potential risks of being penalized, which are quite high. Nevertheless, the large community of users keeps on sharing media content, often long after they've downloaded the files, so others can download them as well.[25] Their motive is not only sustainability, but also the joy of participative activity.[26] Even though dispersed and anonymous, these users form a collective with a strong membership commitment and a clear perception of belonging and solidarity.[27] In a sense, they sometimes act as "recursive communities," that is, communities whose key purpose is the ongoing reproduction of themselves as a community, trumping any material production in which they may engage.[28] Collaborative media consumption signals a possible shift in civic engagement: a less political, more collective, and freedom-of-information-oriented approach can also manifest itself in actual political parties; the Pirate Party in Germany and Denmark is one such example.[29]

The attractiveness of collaboration stems from various factors, one being the joy of taking on a role previously reserved for a privileged group. Just as we described in chapter 3, the experience of writing an encyclopedia can overturn an existing hierarchy; instead of relying on formal authorities, Wikipedia editors need only credible sources, a set of rules, and a little bit of internal "street cred." Another appeal, however, is the participation in a collaborative creative process of both production and consumption of media.

According to Axel Bruns, the proponent of the phenomenon called *produsage*, creating a lasting common good is one of the key motivators for participating, especially given the continuously improved quality and value we demonstrated in the case of Wikipedia or Linux.[30] We believe this is not necessarily or always true, especially in the prosumption of collaborative entertainment. Rather, the value is predominantly performative and relies on experiencing the outcome jointly as the creators and audience shape it.

Take Tumblr as an example. With over half a billion monthly visits, it is one of the top 50 most popular websites in the world—a for-profit platform that hosts nearly 400 million microblogs. Because of the way its user interface is designed, Tumblr is particularly suited for pictorial commentary. Participation in the collective consumption of images is made easy even for newcomers, as Tumblr enables users to follow trending posts across blogs. Its rich community life, however, grows deeper, as Tumblr hosts many fandom communities.[31] Publishers of posted content often interact and cooperate with fandom on Tumblr. But this sometimes causes tension concerning the fans' ownership of characters and stories, and what the fans consider permissible in the reusing and remixing of images.[32] Participation in the *remix culture* often goes against the traditional norms of authorship or copyright, as the remixes refer to commonly known images, videos, or visual quotes.[33,34] Intellectual property law does not adequately recognize this emerging new form of art.[35] It is true even of such innocuous remixes as lip dubs.[36] On the other hand, there is solid research to show that remixing actually increases one's skills and proficiency in a given task.[37]

There are many other websites focused on the aggregated publication of creative images or videos, even without a strong fandom. For instance, Imgur, the image-sharing community portal ranked among the top 40 websites in the world, shows on its main page "the most viral images on the internet, sorted by popularity." As a result, it is significantly skewed toward entertainment and jokes. There are many other websites even more specialized in humorous user-generated content, such as 9GAG. Because the consensus decides which images appear on the main page, a strong user community curates the content, engages in internal discussions, supports other members, and posts comments. Community governance partly resembles the model used at more serious peer production sites, but the consumption of media is much more instantaneous and transient. The very act of collaboration in watching, commenting, and publishing content is intrinsic to the participatory culture of these communities, and it may have an impact on the culture's allure.[38] In a turn "from Generation Me to Generation We,"[39] and the rise of political tribalism, prosumers may actually prefer media that creators either produced collaboratively or that appear to the audience as peer-produced, such as instructional or how-to videos on YouTube, the second most popular website in the world. Many collaborative media websites play a substantial role in intermediating production and consumption, however, which partly resembles the approach of platform capitalism. For instance, YouTube steers its viewers into watching particular videos through the use of nontransparent suggestion algorithms based on rankings, user-driven statistics, and undisclosed proprietary scripts. This lack of transparency suggests that the mechanisms may significantly influence the forms of participation, collaboration, and user agency.[40]

Many image-sharing websites also provide simple meme generation tools, making it extremely easy to create instant visual commentary and further encourage the remix culture. According to Google Trends, "memes" became a more popular search term than "Jesus" in 2016.[41] There are at least two leading types of memes. The first is *visual macro*, a remix made from combining a known picture with novel comments and quotes. The second is *reaction Photoshop*, which relies on removing a recognizable character from its original picture context and putting it in incongruous circumstances or re-appropriating symbols by framing them in a different setting.[42] Both embody an emerging form of art, with its own fashions, cultural heritage, and norms. Internet memes can only exist as a part of a larger whole, as their meaning surfaces in relation to everything else. As a result, they are extremely collaborative.

In 2017, the LG appliance electronics company surveyed 2,000 British adults.[43] The meme they chose as their all-time favorite turned out to be "The Condescending Wonka." The image, based on a scene from the 1971 musical film *Willy Wonka & the Chocolate Factory*, first emerged as a meme in 2011.[44] Since then it has been used in endless

Figure 2 Gene Wilder as Willy Wonka, in a meme appropriated 40 years after he played the role in a hit film. Source: knowyourmeme.com

remixes, typically to convey sarcasm and a patronizing attitude (see figure 2).

But online meme creation and commentary do more than just entertain. Memes can play an important role in new forms of social activism. Take the collective laugh that ensued by repurposing the US presidential candidate Mitt Romney's faux pas "binders full of women," a phrase he used to describe the large number of female job applicants his administration considered hiring several years earlier, while he was a Massachusetts governor. It shows how viral memes on platforms such as Facebook, Tumblr, 9GAG, or even in reviews posted on Amazon.com, can facilitate new ways of expressing political agency and feminist cultural critique.[45]

The Meme Political

The English-language community known by many as the cradle of memes, trolling, and alterculture is 4chan. Its members enjoy dark and morbid humor, weird porn, and practical jokes. (One such "joke" involved persuading quite a few iPhone owners that microwaving their phones was a viable method for charging them.) The 4chan site gained notoriety for posting a large number of leaked, nude celebrity photos back in 2014. It popularized memes such as lolcats and pedobear (yes, a cartoon bear used to mock human predators with a sexual interest in children). A number of major social actions started on 4chan via successful DDoS (distributed denial of service) attacks on the websites of two anti-piracy organizations, the Motion Picture Association of America and the Recording Industry Association of America. And 4chan circulated as well the rumor that Steve Jobs had a heart attack, which led Apple's stock price to drop significantly in 2008. A hacktivist group that we discuss in chapter 5, the Anonymous, also originated on 4chan. More recently, 4chan has experienced sharper right-wing and alt-right turns, even though its uncompromised critique of social norms has been present in its culture since the beginning, and its users often express disgust for "normies" (mainstream people).[46]

Because of and in spite of all these activities, the 4chan site, organized as an imageboard and operating like a standard forum, ranks roughly in the top 200 websites in the world, attracting more than 20 million unique visitors monthly, with daily posts numbering in the hundreds of thousands. It does not permit registration under a user name, so many 4chan users post anonymously; 4chan deletes posts after several days, a practice that's unusual for a forum. The site has a tremendous external impact given its strong countercultural ethos. Many see 4chan as a collaborative reinforcement of a playful cultural expression of post-irony, politically incorrect jokes, and online anarchism. It is a fascinating example of a community able to establish a strong identity with associated norms despite nearly complete anonymity and ephemerality.[47]

Other communities also rely on online collective media sharing, often with a more resounding impact in the world. For instance, the Occupy Wall Street (OWS) movement initiated a protest in mainstream media with a poster in the Canadian-based magazine *Adbusters*. A ballerina poised on the renowned "Charging Bull" sculpture on Wall Street appeared in the foreground, with protesters partly obscured in a cloud of tear gas behind the image. A red tagline read, "What is our one demand?" The text below it, in white type—"#Occupy Wall Street September 17th. Bring tent"—introduced the movement's now-famous hashtag. But the powerful and memorable slogan "We are the 99%"

first appeared on Tumblr. It started with a request from a 28-year-old New York activist, "Chris," who asked people to "submit their pictures with a hand-written sign explaining how these harsh financial times have been affecting them, have them identify themselves as the 99 percent,' and then write 'occupywallst.org' at the end." In less than a month, more than 100 pictures were posted daily on the blog, and the phrase garnered worldwide recognition. The logic of collaborative remixing, connectivity, and personalizing shared content contributed to the powerful dynamics of the social action.[48]

Similarly, a photo taken on the campus at the University of California, Davis, of a policeman who nonchalantly pepper sprayed a group of OWS protesters sitting on the sidelines, served as a powerful image to reinforce the purpose of the movement. The photo underwent many remixes and countless reaction Photoshops. One of the early examples, posted on Reddit, shows the policeman superimposed on the 1819 painting *Declaration of Independence* (figure 3). The meme and its many variations received enormous coverage, including discussion in the *Washington Post*, *ABC News*, *Gawker*, *BuzzFeed*, *CBS News*, *CNet*, and *Scientific American*.[49]

The use of memes in social movements allows for the networked use of visual political rhetoric. OWS memes specifically "employed populist argument and popular texts, intertwining them into a vibrant polyvocal public

Figure 3 This 2011 meme, based on a viral video of a police officer's use of pepper spray during a campus demonstration, expanded the Occupy movement's original focus on economic inequality to include the basic First Amendment Right to peacefully assemble.

discourse."[50] Memes offer a simple way to participate in a collective conversation, in many ways allowing for a deep, even if an acerbic or sardonic commentary on the nature of human liberties.[51] The new collaborative culture of meme creation and consumption is so pervasive that even active military personnel at war feel compelled to participate.[52]

5

COLLABORATIVE SOCIAL ACTIVISM AND HACKTIVISM

We briefly mentioned collaborative social activism and hacktivism in chapter 4, but these topics deserve special focus. There are unlikely to be other types of collaborative activity that rapidly mobilize both online and offline action to the extent that social activism and hacktivism do. Moreover, both social activism and hacktivism are collaborative to the bone: the success of their actions and movements relies on an ever-increasing number of people who join and take part. Active community is both the means and the goal of social activism and hacktivism. In this chapter we thus discuss the specific culture of hacktivism, as well as the blend of technological and social focus in its collaborative initiatives.

Internet activism is also known as web activism, online activism, digital campaigning, digital activism,

Both social activism and hacktivism are collaborative to the bone: the success of their actions and movements relies on an ever-increasing number of people who join and take part.

online organizing, electronic advocacy, cyberactivism, e-campaigning, and e-activism. Whatever term we choose, it involves the use of electronic communication technologies such as social media platforms, emails, and podcasts, as well as memes and GIF (graphic interchange format) files, to gather people in support of a particular cause. Usually a digital activism campaign gives public voice to collective claims on authority. But the internet is undoubtedly a key resource for independent activists—sometimes also called e-activists—particularly those whose message may counter the mainstream. Listservs like BurmaNet or Freedom News Group, for instance, support the distribution of news that would otherwise be inaccessible in highly censored countries. Internet activists also transfer and send e-petitions to governments or public and private organizations to protest against and urge for positive policy changes in areas ranging from arms trading to animal testing. But no matter what form it takes, internet activism enables efficient collaboration among those engaged. The use of digital tools and platforms for social activism enables faster and more effective communication among diverse citizen movements, and the coordination and delivery of particular information to large and specific audiences.

Sandor Vegh divides online activism into three main categories, even if their boundaries may sometimes blur:[1]

1. **Awareness/advocacy** (e.g., #HiddenHeroes or EarthHour by WWF)

2. **Organization/mobilization** (e.g., the Occupy Wall Street movement).

3. **Action/reaction** (e.g., the #MeToo campaign)

There are other ways of classifying types of online activism, such as by the degree of reliance on the internet or the degree of mobilization in the offline world. Online activism could be conceptualized as either a method of working or a digital place for action.[2] According to such criteria, internet sleuthing or hacking are purely online forms of activism, whereas the Occupy Wall Street movement was only partially online.

Vegh's concept of organization/mobilization may refer to activities that happen solely online, solely offline but organized online, or a combination of online and offline. Mainstream social-networking sites, like Instagram or Twitter, more or less directly make e-activist tools available to their users; cyberactivist communities mainly benefit from social networking sites because they allow communication between groups that are otherwise unable to interact. Nessim Watson writes extensively on cyberactivism.[3,] He focused his 1997 article, "Why We Argue about Virtual Community: A Case Study of the Phish.net Fan Community," on the necessity for communication in online communities: "Without ongoing communication among its participants, a community dissolves."[4] The constant ability to communicate with members of the community not only enriches online community experiences but most of all redefines the word "community."[5] We think this claim still holds true after two decades. Cyberactivism in its diverse forms is a true embodiment of online collaboration. Even though we have broadly written about different types of groups and communities based on collaboration, cyberactivism is one activity that embodies collaboration, and without it would cease to exist. That is unless we talk about hacktivism, a very particular and increasingly important form of cyberactivism.

The Many Faces of Hacktivism

What is hacktivism, really? And how do we define it? Let's start by distinguishing hacktivism as the underground use of technology to promote political causes, whereas collaborative social activism—and the self-tracking we describe in chapter 7—relies on collaboration without implicitly being against "the system." (In some scenarios, however, the lack of this "anti" quality may actually disempower social activists.) Hacking as a form of activism can be carried out through a network of activists like Anonymous[6] and WikiLeaks,[7,8] or by a group of individual activists working in collaboration toward a common goal.

Hacktivism as a term is already fairly controversial, with several meanings applied to different contexts. The word was coined to characterize working toward social change by combining programming skills with critical reflection. But a "hack" may also refer to cybercrime, and thus hacktivism may be used to describe malicious activism that undermines the security of the internet. Hacktivist activities span many political ideals and issues. A hacktivist uses the same tools and techniques as a hacker—distinguishing between the two is increasingly cumbersome—but does so in order to bring attention to a political or social cause. Examples might include leaving an attention-getting message on the homepage of a popular website or conveying an opposing point of view. Another tactic might be launching a denial-of-service attack to disrupt traffic to a particular site. For instance, in 2018 David Chesley Goodyear was sentenced to 26 months in prison for a distributed denial of service (DDoS) against the world's largest astronomy forum, Cloudy Nights, apparently because he was angry about being banned from the website.[9] Attempts to pin down clear distinctions between of "good" and "bad" hackers or hacktivists seem pointless, although some use the terms "black hat," "white hat," and "grey hat" hacking.[10] The academic community avoids such labels, focusing instead on the kind

Hacktivism is the underground use of technology to promote political causes, whereas collaborative social activism relies on collaboration without implicitly being against "the system."

of computing expertise the hackers possess. Nevertheless, some critics of hacking and hacktivism raise important issues when deliberating what counts as a crime, and the debates continue in popular discourse. Bruce Schneier, in *Secrets and Lies*, argues that even if the breacher is not malicious, breaching is a malicious act because it makes data less reliable.[11] Many opponents of this view argue that hacktivism is the equivalent of a protest and should be therefore protected as a form of free speech.[12]

Hacking itself has also been frequently defined as a form of culture jamming, or the manipulation of mass media by artists and activists.[13] In most cases, the intent of culture jamming is "to criticize the media's manipulation of reality, lampoon consumerism, or question corporate power."[14] In its intention, this action constitutes political activism, though it may be viewed as vandalism. On the other hand, some may view all kinds of vandalism as a form of activism, even if the act was not intentionally political.

Whereas the value of social activism is seldom questioned, hacktivism remains controversial—even though people widely share many of the principles it upholds, such as resistance to the abuse of power and freedom of expression. For most of history, in one way or another, people have actively demonstrated against—or for—something that they felt passionately about. Moreover, hacktivists rarely seek financial gains; instead, they want to make a statement. In the columns below we describe various forms of hacktivism. Some are more individual, but many forms solely rely on collaborative efforts, without which they would not be effective. Collaborative efforts in hacktivism usually yield the best results because they have the potential and tendency to go viral.

Table 5.1

Leaning toward individual actions	Leaning toward collaborative actions
Website mirroring is used as a circumvention tool to bypass censorship blocks on websites. It is a technique that copies the content of a censored website and posts it to other domains and subdomains that are not censored.	**Code** in software and websites like WikiLeaks can achieve political purposes. For example, WikiLeaks seeks to "keep governments open," but its actions can be perceived as manipulative.
D0xing publicly exposes the personal information of a target. Information in a "d0xing" post may include a current address, social security number, or date of birth.	**Anti-surveillance efforts** such as TOR, Signal, or even Silk Road are usually used for one-on-one communication but have collaborative features as well.
Anonymous blogging is a method of speaking to a wide audience about human rights or government oppression that utilizes various web tools like free email accounts, IP masking, and blogging software to preserve a high level of anonymity.	**Geo-bombing** is a technique in which citizens of the network—also called "netizens"—add a geo-tag when editing YouTube videos so that the location of the video can be displayed in Google Earth.
	RECAP is software that was written to "liberate US case law" and make it freely available online. The software project takes the form of a distributed document collection and archive.

Some of these collaborations are spontaneous and anonymous, but others are carefully organized by consolidated groups of activists. Of the various known and prominent online hacktivist communities, the Texas-based CDC (Cult of the Dead Cow) is predominantly a media website focused on privacy and access to information as a basic human right.[15] Another community, Access to Information, empowers people to make informed decisions. (Philosophically, the two communities share a lot with the Electronic Frontier Foundation and John Perry Barlow, who in 1996 wrote the "Declaration of Independence for Cyberspace.") Both communities are founded on libertarian notions of free access to information and the idea that governments have no right to stifle free speech on the internet. The CDC is also a leading developer of privacy and security tools, which it offers to the public for free, and it participates in several other groups, including Hactivismo and Ninja Strike Force.

Hactivismo, an international group of hackers, human rights workers, lawyers, and artists actually emerged from the CDC.[16] For its ethical point of departure, it takes the principles enshrined in *The Universal Declaration of Human Rights* and *The International Convention on Civil and Political Rights.*[17,18] Furthermore, Hactivismo supports the free software and open-source movements. An internal debate within both Hactivismo and the CDC continues to address the types of ethical behavior considered acceptable in hacking activities. For instance, the CDC launched a number of campaigns aimed at exposing internet censorship in the People's Republic of China because of the country's highly restricted internet service.[19] Other hacktivist collectives, however, do not get fully behind such initiatives.

Anti-ACTA Movement and the Rise of Anonymous

Another lesser known, but no less interesting, example of a distributed hacktivist movement is the worldwide anti-ACTA (Anti-Counterfeiting Trade Agreement) protest. ACTA, with its aim to establish a legal framework for the more effective control of large-scale intellectual property rights violations, had little visibility from the time it was drafted in 2010 until January 2012 after a series of well-publicized DDoS attacks on government, religious, and corporate websites. Thus the strong resistance to the ACTA came as a surprise, both to politicians and the general public, in countries where the protests took place. But because of its unexpected occurrence and scale, the anti-ACTA protest proved to be very influential.

One of the most sudden hacktivist attacks worldwide happened in Poland as politicians debated whether to sign the ACTA. Massive marches and public demonstrations continued for several days, while teams of hacktivists took the government's websites offline for prolonged periods. Then, after Poland's official declaration to sign the ACTA, a number of Polish government websites shut down because of DDoS attacks, which also brought down the official websites of Poland's president, prime minister, Parliament, the Ministry of Foreign Affairs, and the Ministry of Culture. The groups presumed to be behind the attacks were Anonymous and a loose hacker organization called the Polish Underground. Cyberattacks by Anonymous and the Polish Underground made bold statements directed to the Polish public. On the website of Prime Minister Donald Tusk they filled a webpage with a banner reading "Hacked by the Polish Underground: Stop ACTA."

Anonymous, which originated in 2003 on the imageboard of 4chan,[20,21] a site we discussed in chapter 4, embodies a vision held by many online and offline community users who simultaneously operate as an anarchic, digitized global brain: members see themselves as a collective, working together but remaining, even internally, unrecognized and invisible. Members of Anonymous are known as "Anons" and can frequently be recognized in public wearing Guy Fawkes masks, a reference to the mask-wearing anarchist freedom fighter V in the graphic novel and film *V for Vendetta*.[22] Quite often, Anons use the same symbolic masks on their social media profiles. Beginning in 2008 the Anonymous collective became increasingly associated with collaborative, international cyberactivism.[23,24]

The Anonymous collective embraces the idea of identity as understood in two distinct ways. On the one hand, it applies to self-identification with the core values of Anonymous; on the other hand, it applies to the actual identity of people who participate in the Anonymous group actions. The fact that the actual identity of its members remains undisclosed is closely linked to the structure of the movement, but also to its core identity-building idea: anonymous online resistance against internet censorship and surveillance. The group became known after a series of well-publicized DDoS attacks on government, religious, and corporate websites. The essence of this group, then, embodies two somewhat contradicting values: collaboration and anonymity.

The Anonymous organization resembles a swarm (in computing terms the word means multiple systems functioning as a powerful, flexible engine).[25,26] Taking a computing analogy further, Anonymous works like a black-boxed, deep-learning neural network with visible input (a politically motivated goal) and output (political action), but its internal mechanisms are frequently cryptic, which makes for a curious type of self-emergent organization.[27] As a decentralized online community acting anonymously, it has worked in a coordinated manner toward a loosely agreed-upon goal, creating a culture derived from a fundamental sense of humor. This sensibility, in which "it's not the anesthetic humor that makes days go by easier, it's humor that heightens contradictions," is fittingly described as "lulz" (a corruption of the phonetic plural for LOL, or "laugh out loud.")[28] Over time, the lulz developed as a strategy to address change and the need for change as the sociopolitical agenda of Anonymous gained in importance and visibility. Consider its political involvement in 2008's Project Chanology, which comprised a series of hacks and pranks targeting the Church of Scientology.[29] This was followed by the ACTA activities in 2012 and involvements in several other advocacy cases. Simultaneously, Anonymous undertook more serious protests and direct actions. Later targets of Anonymous hacktivism included government agencies in the United States, Israel, and Tunisia; ISIS; child pornography sites (the action called "Anons against Pedos"); copyright protection agencies; and corporations such as PayPal, MasterCard, and Visa. Moreover, Anons have publicly supported WikiLeaks and the Occupy Wall Street movement. Anonymous has undertaken protests and other actions in retaliation against the anti-piracy campaigns by motion picture and recording industry trade associations.

What we observe overall in Anonymous is a particular dynamic of organizing in which individuals form a collective to support its values, but do not necessarily become its members in any binding way. We usually think of *collaboration* as a process that occurs when two or more people, organizations, or institutions work together to realize a goal—a process that typically requires leadership.[30] But Anonymous has no leader or controlling party, and thus it relies on the collective power of its participants to act in such a way that the net effect benefits the group. There is no ranking among participants and no single means of communication, as the collective spreads over many mediums and languages. It is impossible to "join" Anonymous, but "membership" is awarded by the mere wish to participate. Virtually anyone may participate in Anonymous, and unidentified individuals sometimes attribute their actions to Anonymous.

Supporters of Anonymous have called the group members "freedom fighters" and digital Robin Hoods, while critics have described them as "a cybermob" and "cyberterrorists."[31,32] It is true—judging from the WikiLeaks exposure of Hillary Clinton's presidential campaign emails, and to DDoS attacks against governments, banks, and other corporations—that in the last few years we've seen hacktivists majorly disrupting cyberspace and the real world. When closely analyzing both self-hacking and the abundant forms of social hacktivism, we must ask: What happens on the internet as a space, in which we may create alternative spheres of communication? It seems like the uncovering of hidden information went into hyperdrive—with groups such as Anonymous, LuxLeaks, WikiLeaks, and DC Leaks shaping the news and affecting global dialogue, while also contributing to undermining trust in digitally stored information.[33,34]

An entirely different aspect of online activism and collaborative society can be seen through the rapid growth of crowdfunding platforms and initiatives. Both endeavors are theoretically meant as vehicles to fund ventures that could more easily reach their target goals by appealing to a greater number of individual financial supporters rather than to fewer traditional investors. These campaigns can thus end up as hubs for attracting like-minded people. Although crowdfunded projects are often delayed or financially unsuccessful, they are nevertheless popular to drive motivation in (and reinforce) community efforts. This may also be a reason why they are equally popular as new, democratized charity models, and as alternative incubators for business startups.[35] Notably, crowdfunding tools allow for civic engagement and collective action without organization-centered approach.[36]

The Future of Activism in the Digital Age

We may think today that hacktivism serves a primarily political purpose. But it is also a social activity with an inherently collaborative character, at least on the level of mobilizing and increasing exchanges between those who wish to contribute to a common goal. Nonetheless, scholars disagree about the influence the internet will have on political participation. Those who suggest that political participation will increase believe the internet will not only have a role in recruiting and communicating with others online but also offer lower-cost modes of participating for those who lack the time or motivation to engage otherwise.

As Zeynep Tufekci claims in *Twitter and Teargas*, her book devoted to the Arab Spring, the ability to organize without organizations can speed things up at a greater scale within shorter timeframes, as there is no need to deal with the complexities of logistics when crowdfunding can do the job.[37] On the other hand, Tufekci notices that all the tedious work performed during the pre-internet era accustomed people to the processes of collective decision-making and helped create the resilience all movements need to survive in the long run. As Tufekci develops her argument, her account of online social movements shows them growing more ambiguous:

> As of 2016, many protest movements, from Egypt to Turkey, appear to be in retreat or dispersal. And not all movements using these digitally fueled strategies are seeking positive social change: terrorist groups such as ISIS and white-supremacist groups in North America and Europe also use digital technologies

> to gather, organize, and to amplify their narrative. Meanwhile, new movements are popping up, from Brazil to Ukraine to Hong Kong, as hopeful communities flood the streets in protests and occupations.

Other critics have agreed with Tufekci's hesitant attitude about the future of online social movements by presenting a variety of different arguments that range from seeing the internet as a natural tool for empowering terrorist groups to recognizing the structural impossibility of mobilizing and pursuing long-term social action online. Some claim that any form of online activism may actually contribute to decreasing engagement in real life. Critics of online hacktivism quite often dub it slacktivism and clicktivism.[38,39,40] Micah M. White argues being politically engaged these days requires little more action or commitment than clicking on a few links. In response to one view that insists surfing the web can change the world, he writes, "Clicktivism is to activism as McDonald's is to a slow-cooked meal. It may look like food, but the life-giving nutrients are long gone." White goes on to argue, "Clicktivism reinforces the fear of standing out from the crowd and taking a strong position. It discourages calling for drastic action. And as such, clicktivism will never breed social revolution. To think that it will is a fallacy. One that is dawning on us."[41]

The activist Ralph Nader has stated that "the Internet doesn't do a very good job of motivating action," and he argues that the US Congress, corporations, and the Pentagon do not necessarily "fear the civic use of the Internet."[42] Ethan Zuckerman talks about "slacktivism" by claiming that the internet has devalued certain currencies of activism.[43] Citizens may "like" an activist group on Facebook, visit a website, or comment on a blog, but fail to engage in political activism beyond the internet, such as volunteering or canvassing. Zuckerman's argument is strongly Western-centric, however, because it discounts the impact online activism can have in authoritarian and repressive contexts. The journalist Courtney C. Radsch argues that even this low level of engagement was an important form of activism for Arab youth because it is a form of free speech that can spark mainstream media coverage. Additionally, Gabriele Coleman notices that many hacktivists actually do gather offline, appreciate the offline contact, build strong communities, and have a great capability to mobilize quickly.[44,45]

In the light of all these contradictions and questions, hacktivism appears as a very complex phenomenon. Today many people—both involved and uninvolved—seem to ask themselves what hacktivism will lead to. New technologies have already provided activists and protesters with a powerful means to spread their message and mobilize global action. Moreover, technological innovations have given protesters access to hacking tools to conduct cyber operations analogous to street protests and sit-ins.[46,47] Surely, hacking was already an undercurrent in US national elections as a result of the widespread adoption of electronic voting methods beginning in 2002.[48] Nevertheless, hacktivism has become increasingly visible in more recent years, not only as a topic in the information technology media and opinion columns of daily mainstream news, but also in public policy, governance integrity, and national defense debates in the US Congress.

Hacktivism challenges international affairs not only because it goes beyond borders but also because it has become an important instrument of power.[49] In the national realm, hacktivism slowly grew stronger, but for a very long time and, very often, it did so unnoticeably. European hacking was in fact connected with activism since at least 1984 and the so-called Chaos Communication Congress, the largest hacking/security event in Europe, even though German anarchists organized it.[50] Historically hacking had root in activism and was oriented toward empowering grassroots communities in defending their vital interests and common goals. Currently, however, in some cases a mutated form of "capitalist" hacktivism relies on stealing goods and financial assets from internet users.

After all the recent global political turbulences, large-scale data leaks, and weakened democratic procedures, hacking for a cause is now in the position to explode into a complex set of state and local government challenges—and "hackers with a cause" will shape the global dialogue on everything from international relations to financial reporting to local politics in the same way that protesters shaped matters like civil rights and climate change in the past.

In addition, another intriguing phenomenon comes into view: in many ways, hacktivism proves to be the same strong weapon of manipulation used by the political systems it targets. As we write this book, controversy regarding the 2016 US presidential election still remains unresolved, despite strong links to intervention by Russian hackers. The influence and spread of fake news in this context exemplifies an activity that is hard to classify but can be placed within the range of cyberactivism and cyberterrorism, or even cyberwar. The impact of social media's vulnerability and compliance in these attacks has been challenged in the US Senate.[51] We return to this issue in the last chapter.

To make the matter even more complex, hacking has become a tool to serve not only activists but also governments that want to suppress protests. The Milan-based IT company Hacking Team has sold their computer exploitation products to many governments that suppress human rights, from Iran and Saudi Arabia to the United States and Poland.[52]

Despite all the complexities, the primary purpose behind hacktivism is hacking for a cause instead of solely as a means of civil disobedience. Referring to Tufekci's pro and con scenarios in which protesting online and offline lead nowhere, Nathan Heller commented in a *New Yorker* article from the August 21, 2017, issue: "Smartphones and social media are supposed to have made organizing easier, and activists today speak more about numbers and reach than about lasting results. Is protest a productive use of our political attention? Or is it just a bit of social theatre we perform to make ourselves feel virtuous, useful, and in the right?"[53]

Hacktivism has many different faces. Quite often actions described as examples of hacktivism are meticulous and well considered; they do not resemble a one-time spontaneous endeavor, but a series of carefully planned projects. Others, however, are much less organized.[54] And, in that sense, they do not lack the substance that Tufekci and Heller claim are missing in the recent protest movements. This does not mean that the primary rules of hacktivist ethics have not been shaken by recent global turbulences and their consequences; they have. The problem is not even related to the means by which hacktivists operate—that is, using illegal resources to gather material and information important to them or to disseminate their message—but rather is related to the goals. "Hacking a win is a question of principle," Virginia Heffernan wrote in *Wired* in January 2018. When hacking is all about outsmarting the system and finding its flaws, it does not necessarily correspond with something that society would perceive as a greater good. As P. S. Ryan explained it, summarizing the preface of Steven Levy's 1984 book *Hackers*, the general tenets or principles of hacker ethic include sharing, openness, decentralization, free access to computers, and world improvement, but especially the upholding of democracy and the fundamental laws most live by as a society.[55] Moreover, these rules frequently incorporate unlimited and full access to computers, and to anything that might teach us how the world works.[56] Thus, perhaps the most promising way of thinking about the future of hacktivism is to picture it as a sandbox of possibilities that retains freedom and a desired degree of online anonymity, supplemented by the ethos of openness and voluntary cooperation for a good cause. In the future, hacktivist ethics may actually be very important for younger generations to prevent surveillance, build communities, and gather to achieve goals. The more collaborative, inclusive, and clear about its goals the coming generations of hacktivists can be, the better.

6

COLLABORATIVE KNOWLEDGE CREATION

The monopoly on scientific knowledge rests on shaky ground. As the immense success of Wikipedia shows, academics no longer are the only ones who collect and disseminate knowledge. Along with the growing trend toward amateur content generation, a phenomenon we've discussed in previous chapters, comes the general public's wavering trust in science and scientific institutions. Initiatives such as Sci-Hub and LibGen (Library Genesis), which now allow free access to scholarly books, and to journals that once required a subscription, are a sign that scholars themselves contest the traditional system of knowledge dissemination—but these online sites do so without observing copyright law. Even though academic piracy will unlikely become mainstream, the disregard for intellectual property rights in the sciences is already pushing the boundaries of perceived fairness as the distribution of scholarly work moves more toward open access.

The very same technologies that have made it possible for the collaborative society to emerge now allow people to organize into communities that challenge the traditional methods of scientific discovery, and even the scientific method itself; they aim not only to codify and distribute knowledge but also to *create* it. Some of these communities extend into the real world and affect very real issues such as public health.

We are now experiencing a participatory turn in relation to science.[1] Thus, we must closely observe these shifts and changes, and examine their root causes.

For hundreds of years academia relied on the professoriate, a highly privileged and exclusive class who were legitimized in society's eyes to create science. Most people thought of knowledge production as a traditionally hierarchical process. And yet, as the scholar I. Bernard Cohen has shown, the current antagonism regarding science is not new.[2] Now, however, we observe several additional phenomena that come into play to disturb the hierarchy:

- Distrust in formal expertise and traditional sources
- Advancing democratization of knowledge distribution, and automating and digitizing knowledge
- Further democratization of collaborative knowledge creation and tools for its coordinated development

The very technologies that made it possible for collaborative society to emerge now allow people to organize into communities that challenge traditional methods of scientific discovery, even the scientific method itself.

Distrust in Academia

Let's examine in greater detail why academia and science have lost much of the trust they once had, and why non-scientific sources seem to be gaining credibility. The first of these reasons is quite basic and understandable: scientific knowledge is likely to be complicated, nuanced, and difficult to grasp.

The second reason, more complex and multilayered, concerns our attitudes about the credibility of knowledge and the research that informs it. Today we face an unprecedented growth in the distrust in facts as a whole.[3] An outright hostility to experts and the existing, dominant modes of knowledge hierarchies is on the rise, and more and more people share anti-intellectual sentiments.[4] Suspicion toward large media corporations has seeped into the realm of science and undermined the very foundations of Enlightenment premises, causing a major epistemic crisis.[5]

We now live in a post-truth world where just being right is not enough.[6] This applies also to topics that are clearly quite scientific but had not been considered part of the academic domain by the general public, even before the trust crisis deepened. Take dieting, for example. Despite the substantial body of research available about human physiology, the most popular diets are seldom research-driven. In fact, many diets contradict what modern science has widely ascertained, and they can often prove downright harmful. Most of these diets are built around easy concepts.[7] Eat only meat! Eat in harmony with your zodiac sign! Don't eat carbs! Eat only carbs! Even concepts as radical as chocolate-only or steak-and-eggs-only diets exist in the popular discourse. While solid proofs of their effectiveness and long-term harmlessness are still pending, these diets are much easier to grasp and follow than the well-balanced versions based on contemporary academic knowledge.

The abstruse nature of the scientific message is not the only problem, though. For a surprisingly large number of topics that many people care about, a simple lack of research is an issue, and rigorously data-driven conclusions are difficult to draw. For instance, while some drugs have structured procedures for testing and are subject to relatively thorough studies, there are myriad substances about which very little or nothing is known. The nature of the academic method itself often makes it difficult to determine how a single component affects us overall; even when it comes to the oldest alcoholic beverages humans have enjoyed, like wine or beer.

But people expect answers and solutions, and in the absence of a clear response from science, they turn to those who give the clearest message: the advertisers. Take the market for dietary supplements. Clinical data to prove their overall usefulness is still lacking.[8] Moreover, according to a 2015 investigation by the New York Attorney General's office, only 20 percent of herbal supplements from Walmart, Target, Walgreens, and GNC stores actually contained what they claimed. Despite the lack of data to support the efficacy of some supplements and the discovery that the majority of them on the shelves are "fake," the supplements market is worth nearly $20 billion in the United States alone. This phenomenon shows that we are willing to disregard academic sources that appear bleak, indecisive, and nonconcrete, especially when faced with a profession of evangelists who compete for consumer attention in order to sell their products.

The third reason for distrust in academia is more disturbing. We scholars have failed to descend from our ivory towers, and as a result we have successfully ignored the wake-up calls to be more practice-oriented. Even worse, in quite recent history, we have acted unethically in certain circumstances or published blatantly wrong results. For instance, during the infamous 1932–1972 Tuskegee Syphilis Study, doctors tracked the development of the disease under the false pretense of providing treatment for 400 low-income syphilitic African American men; this led to a major—though deserved—breach of public trust in research ethics, and also to the emergence of control procedures, such as institutional review boards. Similarly, the emergence of "junk science" in the form of corporate-sponsored pseudo-academic studies with a clearly commercial agenda, such as how tobacco affects our health, further undermined the social trust in scholarly rigor.[9] As a result, the general reliance on professional institutional authority decreases while informal meritocracy rises.

New and easier ways to access new and emerging sources of information contribute to the fourth reason for the decline in our trust of scientific sources. Such convenience is closely connected to the erosion of institutional safeguards in the distribution of knowledge and the proliferation of fake news.[10] According to the US National Science Foundation report from 2012, more than 60 percent of Americans who seek scientific information on a given matter turn to the internet and peer-produced knowledge first, while only 12 percent still use the online versions of newspapers or magazines, the traditional media formats.[11] During the same period in the EU, 67.5 percent of citizens regularly used the internet, and 54 percent of them sought information on health and general knowledge.[12] In fact, recent studies confirm that the internet has become the main source of scientific information for the majority of the global North population.[13] This means that the traditional sources of information have become far less popular, although they still are—most of the time—much better verified! People generally prefer ease of access to the assurance of quality. Apparently, Gresham's law—"bad money drives out good" applies to knowledge as well, and we can say that "lies spread faster than truth."[14] Admittedly, though, nontraditional knowledge sources may be more up to date.

Collaborative Knowledge Distribution

As we noted at the start of this chapter, the decreased trust in science coincides with emerging models of knowledge distribution that are more egalitarian and collaborative, and are supported by new technologies. In fact, the collaborative aspect of these models may contribute to their appeal in fields across the board, not only in science. The phenomenon of WikiLeaks or the support for Edward Snowden—the whistleblower who released enormous amounts of classified data gathered from the US National Security Agency—shows that citizens increasingly demand awareness and a high level of transparency as well as easy access to information. Moreover, there's a growing interest in a-hierarchical models of knowledge production that oppose the modern state-controlled model of university-fostered formal expertise.

Consider again the example of Wikipedia, arguably the biggest collaborative project in the history of humankind. It successfully runs the largest online encyclopedia and other knowledge repositories, with quality rivaling other professionally developed sources. Yet Wikipedia relies entirely on nonexpert contributions. In fact, many contributors frown upon claiming formal authority when they discuss how the platform should operate. This distancing from authority is part of Wikipedia's open collaboration model, which, as we've shown previously, relies on a transfer of trust from people to procedures; that is, from individual authority to the belief that as long as we follow these procedures, the results must be right.

The trend to participate in knowledge replication, distribution, and active usage spreads widely and goes beyond the phenomenon of peer production. Even in the areas of knowledge that seemingly require very high qualifications and professional training, such as medicine, people turn to the internet to self-diagnose. Based on the first Health Information National Trends Survey in 2005, scholars observed that the possibility to self-diagnose causes "a tectonic shift in the ways in which patients consume health and medical information, with more patients looking for information online before talking with their physicians."[15] "Doctor Google" is increasingly becoming the physician of choice for many members of society, as convenience often trumps quality. Meanwhile, the trust in the knowledge possessed by its traditionally exclusive bearers—representatives of the medical profession—steadily decreases. Those patients also occasionally challenge physicians' interpretations, sometimes with impressive results, especially when it comes to rare diseases.[16] For instance, having turned in frustration to searching online because their doctors were taking too long to come up with an answer, parents in some cases correctly diagnosed lysosomal storage disorder in their children based on information they gleaned from the internet.[17]

Thus we observe that universal access to knowledge and the popular drive to collaborate in sharing it have inverted traditional knowledge distribution models. Contributing greatly to this result is the fact that official medical institutions handle some inherent contradictions poorly.[18] One of the side effects of the turn to collaborative society is the decoupling of knowledge from expertise.[19]

Unfortunately, the lack of expert control over the distribution of knowledge results in cherry picking; and scientists, who are trained to favor a holistic methodology, rarely appreciate a selective approach to results. Take Andrew Wakefield's infamous study published in the prestigious medical journal *Lancet*, which showed that the measles-mumps-rubella vaccine could cause autism. Scholars have since proven this research wrong beyond any doubt and revealed Wakefield's practices to be dishonest, which caused the editors to retract the article and Wakefield himself to be removed from the UK medical register for deliberate falsification. And yet from time to time the anti-vaccine movement persists in citing his study.[20]

The trend to participate in knowledge replication, distribution, and active usage spreads widely, even to areas of knowledge that seemingly require very high qualifications and professional training, such as medicine.

The clash between the traditional mode of knowledge production (and dissemination) and the more experimental, a-hierarchical modes signifies a stronger turn: the next stage in the never-ending process of distinguishing science from nonscience. As knowledge distribution and application by nonexperts grows, we increasingly see that nonprofessionals collaboratively *create* knowledge even in such demanding fields as medicine. For instance, studies show fruitful results from the participation of patients and their families during clinical research, in which their roles evolved from "research objects" to "colleagues."[21,22] This proves again that the collaborative process occurs at the very heart of the scientific community. Experts who have successfully cooperated with participative research communities and enthusiasts have become the avant-garde of participatory knowledge production, not just its distribution.

In the case of medicine, we can partially explain this phenomenon: it's natural for people to want a second opinion and seek out additional information about a health issue in which they have very high stakes; that goes hand in hand with not having to question experts, as long as the information gathered does not strongly contradict what they've received from other sources.

But as we've seen, the trend by which trust transitions from experts to crowds takes place in less specialized contexts. The popularity of the internet increases the general public's trust in information online, but the need to know where the knowledge comes from still remains powerful.[23] Just having the ability to check the source, which the internet makes possible, can provide a feeling of control in an otherwise uncertain situation.

The increasing amount of accessible knowledge can certainly minimize our fear of not knowing enough. But it may also exacerbate our over-reliance on alternative knowledge and deepen the distrust in the established information sources, which may then lead to the perception that inconvenient information is being hidden from the public. Examples of alternative and a-hierarchical methods of knowledge distribution online abound. People who actively practice sports, for instance, frequently exchange alternative medical theories on online forums, which contradict contemporary scientific knowledge. Also, social media sites like Facebook and Twitter have become powerful platforms for information-sharing communities. Many bottom-up authorities on health and nutrition, such as Vani Hari (aka the "Food Babe"), use social media as the main venue for their alterscientific knowledge and resistance activism against the "Big Industries" (Big Pharma, Big Food, Big Oil). The liberation of knowledge sources has resulted in the unprecedented emancipation of users. In practical terms, as the subtitle of David Weinberger's book says, "the facts aren't the facts, experts are everywhere, and the smartest person in the room is the room."[24]

Citizen Science

As the emerging method of generating and distributing knowledge challenges the established model and the traditional authority of academia, the concept of hacking, in this case related to science, again becomes important.[25] The transition from a top-down to a participative knowledge distribution model simultaneously occurs with a general trend to co-create and collaborate in other industries, as we also have seen in previous chapters. There are even social movements that aim at advanced do-it-yourself (DIY) science, democratizing access to research and experimentation in fields as specialized as genome editing, molecular and synthetic biology, or self-made electrode brain stimulation.[26,27,28] Unlike Wikipedia and other knowledge (re)distribution initiatives, DIY movements aim also at creating knowledge itself. In the emerging form of knowledge co-creation in technoscience, commonly referred to as "citizen science," the phenomenon of research by nonprofessional scientists has become increasingly popular: it also includes DIY biology/biohacking movements. Members of these movements study life sciences or apply scientific knowledge outside of conventional means in order to modify traditional practices or provide alternatives to products already in place as market-mediated solutions or sanctioned by mainstream academic centers. Among bio- and lifehacking projects—the latter are intended to improve and manage aspects of our daily activities more efficiently—we can find quirky applications for very specific markets: foot-implanted sensors that receive seismic data in real time, vibrators implanted under a man's pubic bone at the base of the penis, or bioluminescent plants and animals. But biohackers also engage in weightier quasi-scientific endeavors, like cheaper versions of drugs and open-source plasmids that let tinkerers insert any gene into human cells. In its practice, biohacking is both inclusive and participatory, with many innovations coming from professionally trained biologists who nevertheless need to act outside of the traditional system.[29] Interestingly though, existing technological platforms enable them to collaborate and act against existing formal hierarchies and channels.

The biohacking movement proclaims resistance to more classical structures of science production and distribution; in terms of its political scaffolding, it is very much linked with hacktivism as we described it in chapter 5. Individuals who do not necessarily possess the traditionally demanded credentials to participate in constructing scientific facts can form networked groups to collaborate (both online and in the physical spaces of laboratories) and experiment with the possibilities of synthetic biology: in 2014, there were around 50 DIY biology labs worldwide.[30] In a similar vein, the growth of small workshops open to the public that support digital manufacturing, called Fab

Labs (fabrication laboratories), shows a clear need for local, physical, collective production of goods and knowledge.[31] In the online world, platforms such as SciStarter.com provide access and connections to support thousands of academic projects that require communal efforts, from recording frog-mating sounds to collecting and sharing data to protect local waterways.

The DIY biology movement has a deeply democratizing agenda not only in an abstract, philosophical sense. It seeks creative workarounds to increase the affordability and accessibility of innovation that can have enormous consequences, for instance in farming.[32] Thus what defines this attempt is not the rejection of scientific procedure, but rather its interception and extensive use in different institutional settings and with different goals. When engineers experiment with "nature" in order to sustain society defined by the need for constant growth, DIY movements experiment with the concept of society by changing relations between individuals, groups, raw materials, and working devices. Some of these DIY attempts receive blatant criticism from the academic community as being either nonsensical or outright dangerous.[33,34] In response to the latter claim, a growing body of scholars has noticed that just having the access to biological material and digital information about pathogens, for instance, does not suffice to pose an actual terrorist risk to the society; these scholars stress instead that tacit knowledge is key in successful experimentation, and in fact, the DIY movement fosters interest in the mainstream science.[35,36] Studies have likewise revealed that the public and academic fear associated with DIY biology is misplaced.[37] In the United States, even the FBI has successfully cooperated with DIY biology communities.[38] We might think of citizen science as evidence-based academic activism, which typically transcends the online world, even though the internet serves as the necessary enabler of collaboration. Significantly, DIY and citizen science communities often also use nonprofit shared workspaces called "hackerspaces," which manifest peer production principles and governance in the physical space.[39] Hackerspaces serve as trading posts for people with different academic traditions, skills, and experiences.[40]

In many cases, such as smog detection in Poland or radiation detection in Japan, citizen science movements that started online successfully moved offline to stimulate not only local government reactions but also the involvement of academic and business communities, and thus they have had a supportive but supplementary role in relation to the scientific mainstream.[41] Similarly, anti-shale gas extraction groups have been able to coordinate both online and offline activities to exert pressure on local municipalities.[42] Citizen science takes on the risks of the bold experimentation that is decreasingly possible in traditional laboratories, owing in part to the advancing commercialization of science and in part to the ritualized process of climbing the career ladder (for instance by sticking to publishing safe articles about standardized studies).[43] But the citizen science movement is countercultural only in the sense of challenging the established hierarchies and structures of knowledge creation, and not in questioning the science itself. In fact, the goal of many citizen science initiatives is to produce knowledge that will be respected and valid just as much as the traditional forms, although often with more focus on an overall social impact and social responsibility.[44,45] As a result, the institutional hegemony in biosciences is increasingly challenged by activists and informed consumers, although the market and corporate resistance is still high.[46] Yet none of the successes of citizen science is even close to that of Wikipedia. At the same time, we must emphasize that abuse in citizen science takes place just as much as it does in peer production and collaborative economy projects, given that citizen science movements sometimes function as a source of unpaid labor, exploited by semi- or openly commercial ventures.[47]

Alterscience

Citizen science movements generally support and build on the body of science developed so far, contesting only some of the formal structures and the system of knowledge hierarchies. This is not so in the case of alterscience and antiscience communities. Although they also rely on a collaborative model and question the formal expertise and knowledge hierarchy, they often build their views on the world in opposition to what mainstream science has to offer or by a very selective use of information. Citizen science movements, for example, value diversity of views and treat dominant scientific narratives as a dictatorship.[48] Disappointed by perceived historic failures of science and medicine, they express doubt in scholarly authorities, supported by the decaying belief in scientific objectivism.[49] The fear of the market-driven logic of scientific discoveries and technical innovations of the Big Industries creates yet another reason for distrust. But being full of tin foil–hat paranoia and prone to conspiracy theories, alterscience communities still need to refer to authority figures of their own, either by including scholarly research from outside the mainstream academic resources or by differently interpreting commonly acknowledged results of scientific studies.[50]

The reference network of alterscience communities consists of some diffused but interconnected online social actors: independent journalists and activists; engaged scientists, sometimes with degrees in unrelated fields; and even general practitioners. The communities occasionally flirt with the consumer rights movement and environmental NGOs. They organize through Twitter or on "alternative media" platforms like NaturalNews.com, Global Revolution TV, TruthWiki.org, TheAntiMedia.org, or the GoodGopher browser. It is worth noting that addressing alterscience views by posting on social media sites—for instance, confronting vaccine skeptics—can amplify a debate and encourage Twitter bots and Russian trolls, which have been shown to foster discord.[51]

Wikipedia, Doctor Google, the alterscience movement, and citizen science all have something in common: they rely on the active use, combination, production, and distribution of knowledge. Since the year 2000 or so, we've observed a major trend: the production of information increasingly accompanies knowledge consumption, as in the case of the remix culture we discussed in previous chapters.

We have also witnessed the rise of the "confessional society," a phenomenon described by the Polish-born sociologist and philosopher Zygmunt Bauman, which relies on the willingness of people in general to share their daily life and observations with strangers via the use of technology.[52]

The reasons why people trust online resources change, however, and we attribute this phenomenon to the fact that the internet steadily accustoms users to reading information of differing quality; thus, it develops their defense mechanisms and prevents them from falling for misleading information. And yet disinformation (intended purposely to mislead) is a rapidly growing problem. In fact,

Donald Trump's 2016 US election victory and the UK's Brexit campaign success that same year are both partly attributable to an avalanche of fake news and skillful trolling.

The general information overflow combined with the increase in fake news presence has increased the importance of our proficient use of search engines along with the ability to access the relevant filter bubbles and to prioritize alternate knowledge sources over traditional textbook knowledge.[53] Since traditional information sources slowly lose trust while fake news propagation thrives, people succumb to new forms of tribalism: they trust the online communities that they know and to which they belong.

But active knowledge communities also form for no apparent practical, applied reasons at all. Amateur science, biohacking, the cyborgization of bodies, or alternative medicine and nutrition practices—all of these phenomena aim at increasing participant influence on their environment in general.[54,55,56] As Richard Sennett observes, the dispersed capacity of people to alter their material condition and have agency plays a crucial role in enacting a truly civic democracy not run by expert elites.[57] Subverting the dominant hierarchy of knowledge has been possible thanks to new technologies, as they allow for the dispersed generation, storage, and distribution of knowledge. But they signal a much bigger change: from an authoritarian knowledge creation and dissemination pyramid to a collaborative networked system.

7

COLLABORATIVE GADGETS

In this chapter we explore the complex role that ubiquitous technologies play in forming collaborative communities and practices. Although it is obvious now that technology is a foundation and enabler of online collaboration, wearable technologies have a distinctive function in this landscape, significantly different from the general IT infrastructure. Cloud technology, mobile technology, collaborative applications, the Internet of Things, Big Data analytics, machine learning, and narrowly specialized AI—all of these phenomena promise to radically enhance the way we work and interact by allowing us users to optimize our time, facilitate interactions, achieve visibility, and drive us toward constant collaboration. Taking a closer look at how technology steers societal change, particularly in the context of collaboration, we focus on tracking technologies that are currently being rethought and reshaped:

from self-tracking—intended to monitor simple and easily quantifiable activities—we move on to examine more collaborative and sophisticated forms of tracking the self and others. We aim in this chapter to show how the proliferation of tracking hardware and applications has also allowed new, more complex collaborative aspects of self-quantification to emerge.

Arguably, the development of new tracking technologies has not only altered the ways individual humans think, and how they identify and express themselves, but it has also enabled revolutionary cultural change, sometimes transforming practices that have been central for centuries or even millennia. The development and widespread dissemination of the written word, for instance, to a certain degree diminished the role of oral poetry, once the mainstay of cultural transmission.[1] Even some abilities that defined the perception of intelligence in ancient times—such as memorizing, improvising, and performing thousands of lines of poetry—exceed the present-day cognitive abilities of almost all humans. We are now undergoing a profound transformation sparked by developments and rapid innovations in the field of internet and communication technologies (ICT). But as technology emerges to become the key driver of collaboration, it in some cases becomes its prime enemy.

In earlier chapters we explored the impact of social media on the daily lives of us all, from its empowering effects in bringing online collaborative efforts together, to the negative aspects that result from giving up control over our data, such as the lack of privacy and the fear of surveillance. In the second decade of the 2000s we've seen how digital tools can enable online collaboration in any number of wide-ranging projects, and we've observed how the internet can mobilize online activism.[2,3,4,5] And yet scandals caused by data leaks and breaches of privacy left many people feeling threatened about how the digital transformation of life will increasingly affect them.[6,7,8] Technology has been responsible for major impacts on the workspace and workforce, with recent developments ranging from security to management to wellness applications; in any of these workplace scenarios software can manage group discourse that often enough results in shared understandings, new meanings, and collaborative learning. But as the impact of technology extends further into the future, some tech experts predict that workers in a predominantly digital or gig economy will live their lives according to the rhythms of an app that tells them what to do on the job, based on previous and predictable tasks.[9]

This technology, on the other hand, generates myriad problems and doubts related to the following issues:

- Matters of safety, privacy, and control over data accessibility

- Work inefficiency due to various online distractions
- Addiction to social media and technological gadgets
- Consequences of misuse and the exploitation of data

Much of the new technology depends on cloud computing, using the ever-present internet to provide data systems and services via mobile networks, which in turn make possible lightning-fast interactivity with vast warehouses of data. Alongside the interactive possibilities, however, exist unprecedented opportunities to surveil individuals. Although surveillance is obviously problematic when it comes to our privacy, it can also offer novel cognitive and collaborative affordances as the technology becomes more attuned to our cognitive profiles. The intimate nature of the seemingly ubiquitous cloud technology that is now packaged with our smartphones and wearable technologies means that ICT is ever more closely enmeshed with our organic cognitive faculties, such as imagination, perception, thinking, judgment, and memory, and the collaborative nature of our tasks. These new technological devices do not necessarily replace our organic operations, but they certainly augment, extend, and sometimes diminish our capabilities to perform them. This is relevant in terms of megapixel cameras, smartphones, iPods, and iPads—and especially wearable devices, which are the main object of our consideration in this chapter.[10,11]

Wearables and Their Collaborative Appeal

Today the tensions and ambiguous influences of technology are prone to materialize in tracking technology embedded in wearable devices, which often go hand-in-hand with smartphones. Tracking—that is, monitoring, measuring, and recording elements of one's body and life as a form of *self-improvement* or *self-reflection*, but also as an improvement and reflection of *others*—has been discussed and practiced since ancient times. But the introduction of digital technologies, especially biosensing, has opened a different level of debate and created an environment for communities focused on tracking activities.[12] Personal data analytics as we know them today began with lifelogging in the 1980s.[13] The proliferation of mobile digital devices enabled lifelogging tools to break out of research labs and move to the hands of the masses. Recently we have experienced and observed an explosion of tracking hardware and applications along with the emergence of new, collaborative aspects of self-quantification, in which the focal point moves from the individual to the virtual community. In this chapter we trace this change and examine its potential ramifications.

The introduction of digital technologies, especially biosensing, has opened a different level of debate and created an environment for communities in which to organize around tracking activities. "Trackers," "wearable technologies," and "wearable devices" are terms that describe electronics and computers integrated into accessories and clothing that we wear on our bodies in the form of jewelry, glasses, watches, and headbands. Recently these devices have begun to significantly affect the areas of fitness, health, and medicine, even though the concept of a wearable is hardly new.[14] For instance, the seventeenth-century Chinese abacus ring allowed bean counters and other "accountants" to perform mathematical tasks by moving tiny beads along nine rows.[15] And in the early 1960s, long before the development of the Internet of Things, Edward O. Thorpe and Claude Shannon created an electronics- and wire-stuffed "smart box" that could fairly accurately predict the end position of the roulette ball; it was meant to be worn strapped around the waist, with input and output made possible (respectively) via a tap of the wearer's shoe and an audible earpiece.[16] This gadget and others of its era, however, did not reach a wide audience, and their availability was limited to experimentation.

The most popular functionalities of wearables include physiological data measurement and biofeedback during sports.[17] Such devices, which focus on displaying data, are called "passive technologies."[18] They are different from active devices, which not only measure a user's behavior or activity but also react to it, for example by applying electric current stimulation. Wearables available today range from personal activity trackers and step trackers to devices that log food intake, monitor heart rate, gauge skin temperature and perspiration, and monitor sleep.[19] Popular brand-name devices include Basis, Shine, Withings Pulse, Fitbit Force, Jawbone, Garmin, and Polar Loop, whereas applications like Endomondo, Runkeeper, or Strava function directly via smartphone operating systems.

One of us, namely Dariusz, while delivering a course under the auspices of FIFA (Fédèration Internationale de Football Association), had a chance to discuss sports trackers with professional soccer team managers. One of the biggest benefits for players who wore trackers during all sports activities—including outside of gym and team training contexts—resulted from the managers' ability to confront players about their activity levels with actual data; this is similar to the improvement a GPS tracker can bring to location self-reporting.[20] Trackers thus have potential as tools for direct, social control and for quickly implementing self-control in users. Even though some users attempt to outmaneuver the system, for example by reporting that a device broke or did not charge, such strategies prove not in their best interest in the long term.

Most wearable technologies synchronize with personal tablets and smartphones, in which dedicated applications store and analyze data, as well as share data with friends and community. Sharing data on social media,

The collaborative component of wearable devices (for the most part) constitutes their usability and appeal.

or at least within a given wearable social network ecosystem (particularly when it comes to medical data), is usually a default option: without the possibility of praising oneself for the accomplishment ("I ran 5 miles today!") and making comparisons with others, the functionality of such applications would greatly diminish. The collaborative component of wearable devices (for the most part) constitutes their usability and appeal.[21,22]

Quantified Self and Quantified Others

Quantified self-tracking first appeared in the health sector and then moved to recreational sports and wellness, where the technology became mainstream. Biomarker testing and health metric tracking was once an expensive one-off process traditionally ordered by physicians for patients in response to specific medical risks. Two of the biggest applications of doctor-driven health metric tracking have been cardiac monitoring and telemedicine (remote diagnosis), in which implantable, worn, or handheld devices wirelessly transmit data to medical professionals.[23,24] A number of different initiatives now attempt to facilitate collaborative health monitoring, including the emergence of internet-based social networking communities and newly available low-cost technologies like genome sequencing and biomonitoring applications and devices.

The Human Genome Project would be impossible without the data sharing of anonymous donors and biomonitoring. Moreover, some of the routine services and monitoring processes related to the collection of patient biodata—conventionally conducted at clinical sites—have been successfully delegated to individual remote monitoring systems outside the clinical environment. This obviously reduced healthcare costs, but also encouraged nonexpert communities to undertake biomonitoring efforts by themselves. Yet another interesting layer of this process concerns how the demand for genetic information to satisfy personal curiosity, augment clinical care, and enable vital research increases the pool of information that data companies can benefit from.[25] In fact, data custodians push for wider access to biomonitoring and genetic testing because they can monetize the data, whether by selling to businesses or other profit-seeking organization. This is not exactly the collaborative data sharing access we were emphasizing before, but it is a great incentive to those that mainstream and enable data sharing.

Quantified Self (QS) is the term that "embodies self-knowledge through self-tracking."[26] QS practices a different approach to the so-called $n = 1$ studies (clinical trials or research in which a single patient is the entire trial, the single case study). In the past, n equaled a "someone else" whose data could be applied to represent the population average, but that has changed dramatically with the advent of personalized tracking where *n* equals "me" (the user). Although most of the commercially available tracking devices primarily measure individual progress indicators, their added value is unveiled precisely when they become tools for collaboration with others. Obviously, these two tracking functions—individual and community—are interdependent and often difficult to separate. Many individuals use wearable devices, but the most engaged usually join already existing self-tracking communities or establish new ones. Several interest groups such as Quantified Self, HomeCamp, DIYgenomics, and PatientsLikeMe have formed since 2008 to explore, brainstorm, and share their self-tracking experiences.

According to the online and offline declarations of QS members, the ultimate goal of these communities is to smoothly integrate the technology of human body tracking with the daily lives of individual users, the goal being to gain and benefit from personal self-knowledge. In this context, the human body becomes the central element of human-computer interaction by moving away from desktop applications toward mobile and wearable ones. The underlying assumption here involves seeing data as an objective resource that can quickly bring visibility and information to a situation and, psychologically, entail an element of empowerment, control, and fun. These communities coined and developed the above idea of $n = 1$—in the dual sense of experimenting with tracking devices on the individual level, and of coming together in health collaboration communities to discuss the discoveries—to make the results less anomalous and more statistically significant. Online data sharing, meetups, and common discoveries related to patterns of achieving well-being and self-diagnosis consolidated and sustained these groups until they gradually started to resemble social movements.

Based on the foundation of group self-tracking practices, an ideological agenda emerged. It included two basic premises: (1) advocating for self-knowledge through numbers, meaning the measurements we gather about ourselves and our activities, and (2) engaging in biocitizenship as a means of resistance when biopolitics acts as a control apparatus exerted over the population as a whole.[27,28] This is a clear reference to Michel Foucault, who devoted much attention to the issues of biopower and biopolitics in his lecture series published as *Society Must Be Defended.*[29] The Quantified Self community often stresses resistance toward biopower, understood as the extension of state power over both the physical and political bodies of a population. The notion of resistance is frequently emphasized while encouraging Quantified Self representatives (known as QSers) to alter the ways in which mainstream devices operate and perform, and to reverse-engineer them.

Data Sharing as a Collaborative Practice

The members of tracking communities (like QSers) advocate widespread anonymous data usage and sharing, as well as data transparency. Within the Quantified Self movement, for instance, ongoing debates related to the types of personal data gathered and known about each of us today make the subject of open transparency very tricky and troublesome to some.[30] These data may include (but are not limited to) details about an individual's time spent online, physical location, expenditures, credit history, net worth, diet, biomarkers, DNA, driving patterns, and criminal behavior.

When it comes to how data are gathered, we mainly concentrate in this chapter on frequently used, commercially produced devices, but it is important to remember that many QS and biohacking communities rely on self-made (DIY) wearable gadgets or reverse-engineered devices. This reflects part of the QS ethos: to make the best use out of wearable devices, not necessarily by adhering to the functions designed by the producers, but rather by transforming the devices to meet the needs of individual community users.

QS research also involves data exchange and data sharing for the greater good. John Wilbanks, an avid QSer who works for the nonprofit Sage Bionetworks, is a great example of someone who actively advocates the open sharing of health data for research. Wilbanks became widely known in 2012 for leading a "We the People" petition in the United States to get the Obama administration to support free access to taxpayer-funded research data.[31] He also worked on the Consent to Research project, which provides a platform for people to donate their health data to advance medicine and deepen scientific research. Taking into account that health data is subject to restrictions and expensive to conduct, this project provided a viable way to positively benefit medicine and patients at large.

Quantified Self encourages more proactive self-management when it comes to healthcare by engaging with health-related communities that operate in a playful, gamified manner, such as Care Opinion in the United Kingdom and ZorgKaartNederland in the Netherlands.[32,33] Both platforms offer ratings for healthcare facilities based on clinical performance and the personal experiences and feedback of patients. They use gamification practices—such as personalized videos, links to live animation, and story telling, all of which fall under the umbrella of play—to foster the collecting, collating, and analyzing of minute quantified data, all of which come back to the users in the form of suggestions about how to take better self-care. The platforms use visual cues and icons to link to cross-referenced information so users can discover relevant topics and get feedback for behavior modification.

On this platform gamification certainly plays an important role in making tracking more collaborative.

Beyond healthcare, gamification proponents promise to make the responsibilities of real life more like a game with applications diverse and wide-ranging, from car dashboards that use mini-games, graphics, and other visual feedback to reward reduced fuel consumption to software that allows users to set, track, and achieve financial management goals. The examples abound: websites that reward users who post interesting comments with reputation points (also called upvotes) and recognition; programs that promote healthy eating habits with points; and game consoles geared to fitness and weight-loss, which are guaranteed to get users off the couch. Online app technologies like Nike+ and Mint pledge to make everyday tasks such as exercising and sticking to a budget more enjoyable. And Foursquare promises to make the supposedly pleasurable practice of socializing even more so.

When it comes to the added value we gain from interacting with others via our apps, tracking apps such as Strava and Google Fit exemplify how the gamification of sports can both motivate participation and keep the competitive spirit alive. Game mechanics not only enhance data visualization on the physical tracker and its accompanying website or software, if applicable, but they also encourage users to seek incentives and perks for sharing that data. Today customers can receive health insurance discounts in exchange for their Fitbit data and measured progress in self-care, or they can earn car insurance discounts in exchange for data from NaviExpert and other systems.[34] Now comes a question: How do these scenarios translate into collaboration? The answer is neither simple nor (as yet) clear. On one hand, the collaborative component in these devices is often a major motivator and commitment builder for, let's say, nonprofessional athletes who want to improve their performance in sports. On the other hand, the collaboration itself happens mainly within the device by showing rankings, statistics, and the users' own achievements over time. The feeling of collaboration is (at least partially) supplanted by a data-driven approach to the results of users themselves and of others.

As self-tracking communities were established, users experienced an evolution in tracking hardware and software. As a response to the need for more "ambient intelligence,"[35,36] where intelligent devices can be easily integrated with everyday surroundings, wearable devices became sophisticated technologies to expose user activity otherwise invisible. We've observed this change, from simple measuring and step counting devices that monitor a walk or a run, for instance, to the rise of increasingly sophisticated trackers, including those that measure brain activity. Muse, a brain-sensing headband for training, relaxation, and meditation, is one example, and similar to mind-trackers such as Emotiv. These trackers operate on an entirely different level of interaction with users and introduce ramifications that challenge our previous understanding of the tracker's role. They are unobtrusive, more personalized, and usable on demand (although not always fully functional and reliable, depending on weather conditions, type of activity, and other circumstances). Although these trackers measure and provide feedback about very complex activities—like moods or emotions—they still retain a high degree of portability and are easy to use. Thus people may use them in various spaces, for instance at group meetings, workplaces, or collaborative relaxation sessions (as Muse encourages). What is more, the rise of Big Data analytics and advances in machine learning enable more robust analyses of all users' digital traces combined.[37,38] As the authors of the report titled *Health Wearable Devices in the Big Data Era* argue, "experts envision a not-too-distant future in which health and wellness devices—along with an array of next-generation Internet-connected sensors—will be fully integrated into everyday experiences as people continue to adapt to the now-ubiquitous presence of digital technology in their lives."[39]

Quantified Self and similar communities can clearly empower their members to adopt more proactive attitudes when it comes to their health, well-being, workplace efficiency, workflow, and even creativity; and they can teach members to rely more on auto-analytics and community feedback. Thus, these communities facilitate the rise of new organizational structures built on horizontal frameworks, creating a powerful environment for data exchange and collaboration between the users.

But to fully explore the potential of highly personalized data open to collaboration, we need to consider several issues. As we've shown, self-tracking is increasingly less of an individual action than its name implies. It may take the form of collaborative tracking and discovering, but just as well it may solely involve tracking others—that is, exercising control and surveillance. No doubt we can identify the full gamut of tracking users, from those entirely self-motivated to those subject to commercial exploitation. The change toward even more meticulous and personalized tracking—combined with efforts to create consolidated tracking communities—is of crucial importance: especially so because tracking parameters that correlate best with various processes and the evolution of context-aware systems (such as those that track location, network usage, or even social connections) can introduce a profound change in the way we think about collaboration. They may either empower or disempower the community. In the business world, that could mean they either redefine top-down solutions in corporations and institutions or consolidate them.[40] In the empowering case, employees could benefit from tracking, realize their professional goals, and be a part of an active community that supports them in doing so. Such a flat structure within an organization would be based on trust and mutual benefits from the use of technology. In the disempowering scenario, tracking could be used not only as a tool to pressure employees to achieve goals relevant for the organization but also to surveil and monitor employee activities.

Another matter is the growing deprofessionalization of those who indeed have expertise in a given field, a phenomenon we mentioned in chapter 6. Let's again consider healthcare, a field in which doctors, trainers, and diet specialists currently face off with the virtual support provided by communities of nonexperts. Such communities, armed with tracking devices, take over the trust previously allocated to the expertise of professionals. This new digitized trust results from familiarity built over time with the online community as well as the objective knowledge the device represents.[41] An interesting issue here, although it cannot possibly be resolved at this point in the development and implementation of wearable technologies, is whether in the future the trackers will somehow replace experts themselves or instead become the intermediary layer between the doctor and the patient. Both scenarios mean more responsibility on the patient's part, but only if the tracker becomes the intermediary will we see more collaboration between humans.

Situations where the expert is replaced by the device entail many other complications. First, the devices can both deceive and endanger the users. Wearables may introduce coercive control (entirely peer-driven), from which emerge strong and consolidated communities like Weight Watchers.[42] Their members may believe more in what their trackers say than in expert medical advice. This results from the perception of computers as depersonalized and objective whereas humans are subjective, opinion-driven, and prone to mistakes.[43]

Many experts agree that self-tracking movements can promote an addiction to technological gadgets and data.[44,45] It is true that measurement and tracking tools can drive users into repetitive cycles of performance by challenging them to reproduce or "better" their previous records. In itself, this is not bad. According to data we collected at Quantified Self meetups, various QSers characterized themselves as "downgraded" or "mediocre" without a tracker and experienced depression whenever a tracker was not displaying their progress. Some reported feeling an urge to migrate without any clear purpose to ever-newer tracking software and hardware. Moreover, these QSers perceived other self-tracking communities as elite clubs whose members have better access to each other and generally are better informed than others. This, again, links to the problem of deprofessionalizing experts in the field.

Tracking can lead to many unplanned consequences. For instance, the Fitbit sports activity tracker allows users the option to manually record an activity, measured in an intensity-range from passive to vigorous. In 2011, when the online data from 200 user profiles showed up in Google search results, including their self-recorded "sex-ercise" statistics, the company took immediate steps to hide user activity from the public.[46] Somewhat similarly, data from the popular running tracker Strava inadvertently exposed the location of several secret US military bases in 2018 by publishing the routes of all users, including military personnel.[47] Data from the Apple Health application, installed on the iOS of all iPhones beginning in 2014, became admissible evidence in a murder case in 2018.[48]

Researchers and users frequently think about self-tracking as either good, empowering, and helpful for people resolving common issues—or essentially evil, highly addictive, and enslaving. This dichotomy not only relies on subjective judgment but also leads to massive oversimplification. And it raises a meta-level issue for the future, as advanced versions of trackers may become highly personalized self-management devices. As Elizabeth Pantzar and Mike Shove note, "Once equipped with a heart rate meter, an individual becomes a knowable, calculable and administrable object."[49]

Essentially, trackers and tracking apps are, like many other technologies, producers of the self. Trackers shape our images of ourselves and of others, and thus they change us. The kind of self they shape is one that strives—or is urged to strive—for a precise, measurable perfection. Moreover, trackers may become even more pervasive producers of self than other devices because the nature of their relationship with users is close and built on trust in the device. Hence the device becomes an important channel to connect to ourselves, by displaying data about us that we could not otherwise know, and to others, by data sharing and exchange.

This sense of connection is a vital aspect of collaborative tracking that many frequently neglect. The very nature of our relationship with tracking devices is also one of the crucial reasons why the self- and others-tracking industry has developed so rapidly. But the future relationship between users and their trackers must become less cumbersome. We know from our research and from other authors that, on the individual level, self-tracking—if not addictive—over time frequently turns out to be boring or frustrating.[50,51,52] When self-tracking nests in a community, however, the collaborative aspect enters to make it less so.

The Future of Collaborative Gadgets

Self-hacking and self-tracking share some features with hacktivism but remain less collaborative and more self-oriented. If we understand self-tracking as an attempt to trust a solid database and its associated devices more than the expertise of professionals—such as the doctors, trainers, and diet specialists related to our healthcare examples—we see similarities with the hacktivism movement that expresses fundamental distrust toward the knowledge distribution system. Particularly in the case of diet tracking, users provide responses in many cases via crowdsourcing or collective feedback from other users and professionals.[53] What's more, crowdsourced feedback is an important self-commitment mechanism for helping to lose weight,[54,55] Hacktivism also shares several common features with the DIY movement, particularly biohacking, which advocates for the kind of mental and physical self-enhancement we described in chapter 6. In this case the common "hacking" component indicates the joy experienced from the intellectual challenge of creatively overcoming limitations of software systems and achieving novel and clever outcomes.[56]

Most probably the future of tracking lies in collaborative endeavors, because they allow for the effects of scale and are most likely to receive systemic support from institutions, companies, and gadget producers. But there's another reason for envisioning a collaborative future: if individual tracking becomes monotonous, being part of something larger than oneself can have an empowering and motivating effect. Gadget producers know that too, which is why they strive to bring trackers into the wellness programs of organizations and corporations and frame them as transparent companions of everyone's routine.

The Quantified Self is currently moving toward the concept of quantified others, in which the *n* we mentioned at the beginning of the chapter really means "all of us," a designation that can be understood as a sphere of connected minds (noosphere) and personalized datasets comprising the digital traces of all users. Despite privacy issues, many users still want a benchmark, and want to see themselves in a continuum of general community achievement. Even more important though, on a structural level, is that Big Data analytics (and sometimes classic machine learning algorithms such as regression or classification[57]) require huge amounts of other users' data. So, even if a specific user does not want to share personal results and compare them with others but still wants meaningful statistics, a great deal of (sometimes hidden) collaboration is required with the device that tracks, shares, exchanges, and computes other users' data.

In his 2014 essay "The Planning Machine," about the origins of the Big Data nation, Evgeny Morozov recalled the famous Project Cybersyn established in 1972 by Stafford Beer, invited by Chile's top planners to help guide the country down what its Marxist leader Salvador Allende called "the Chilean road to socialism."[58] Beer was a leading theorist of cybernetics. At the center of Project

Cybersyn—short for “cybernetics synergy”—was the Operations Room, where advisers made cybernetically sound decisions about the economy. “Those seated in the Operations Room,” Morozov explained, “would review critical highlights—helpfully summarized with up and down arrows—from a real-time feed of factory data from around the country.” What many forget is that Beer also designed an important supplement to Cybersyn—Project Cyberfolk—“to track the real-time happiness of the entire Chilean nation in response to decisions made in the Operations Room. Beer built a device that would enable the country’s citizens, from their living rooms, to move a pointer on a voltmeter-like dial that indicated moods ranging from extreme unhappiness to complete bliss.” Morozov called Project Cyberfolk “a “dispatch from the future” as he explained the rapid progression of Big Data as a concept to be reckoned with in the twenty-first century:

> These days, business publications and technology conferences endlessly celebrate real-time dynamic planning, the widespread deployment of tiny but powerful sensors, and, above all, Big Data—an infinitely elastic concept that, according to some inexorable but yet unnamed law of technological progress, packs twice as much ambiguity in the same two words as it did the year before.[59]

This is why, despite the potentially promising collaborative dimension of tracking, there is a concern that it will become even harder to identify the boundaries between self-driven collaboration and blatant commercial exploitation. Perhaps such polarization is artificial and counterproductive because the communality of trackers does not have to be compromised or destroyed by its commercial dimension. But—bearing in mind sensitivity of data and the particular organizing power of tracking devices—we may ask what kind of collaboration happens here. Wearable devices promise us a lot: well-being, increased productivity, and higher quality of life.[60] Currently, most of the collaboration happens without the user, who must only wear the device and produce and collect the data. And the more users collect more data—the better for the producers, but not necessarily for the end users. Thus, it is important for us to ask: Do tracking devices reduce or enhance collaboration? Who is collaborating with whom? What rules govern the relationship between the technological system, the community that uses it, and the individuals who feed it with data for their supposed benefit? Perhaps, in order to achieve the level of collaboration that really empowers individuals and communities, we need strong and savvy communities, capable of establishing rules and representing their interests against device producers. Despite its popularity, Quantified Self is still a minority movement compared to the whole data flow. More collaborative practices could actually help with data accuracy instead of reconciling with data ambiguity. For the future, we need more communities that feel confident working with data and checking the accuracy of the devices they may use. Moreover, these communities need rules regarding who should be legitimately authorized to view the data of others, as well as whose data must be transparent.

8

BEING TOGETHER ONLINE

The popularity of social media sites today can be attributed to many different factors, but sharing information with others is definitely one of their most important features. We nevertheless treat these platforms primarily as places to go to network, interact, and gain exposure, leaving collaboration out of the picture. But let's take a look at social media through a different lens—one that reveals a particular way of being together, and unexpectedly triggers open collaboration and collaborative projects. In this chapter we examine how various platforms that were not designed for collaboration may unexpectedly facilitate collaboration tools, and how being with other people online can introduce users to new skills and experiences that allow collaboration to thrive.

We also discuss in this chapter why some kinds of collaboration—including collaborative knowledge

Platforms that were not designed for collaboration may unexpectedly facilitate collaboration tools, and being with other people online can introduce users to new skills and experiences that allow collaboration to thrive.

production and consumption—have become an essential part of being together when it comes to social media. Even though studies report that teenagers and millennials use social media mainly to share and collect information,[1,2,3,4] while on these sites they can meet with their peers free from adult supervision and get things done: exchange ideas, prepare homework, and organize other projects.[5] Other studies reveal that young people use Facebook and Snapchat not only to stay up to date with peers and to participate in peer culture, but also to engage in political activism.[6,7,8]

Social media designed for one purpose may turn out to serve another purpose, so instances of collaboration may occur not by design but by accident. Altering the main function of a given medium has become a general trend in itself. Ethan Zuckerman describes an example of this type of unintended opportunity in his *cute cat theory of digital activism*, which also relates to our discussion about cyber-activism and hacktivism in chapter 5.[9] He observes that internet users are more likely to surf the web for low-value but viral online content—such at cute cat videos—than to participate in activism, and because of this various means of web censorship can be sidestepped. Lolcats and similar humorous materials can nevertheless carry (and serve as covers for) political messages dispersed among memes or attention-drawing gifs and pictures. In this sense, Facebook, Twitter, or Flickr are very useful platforms to social movement activists, who may (and usually do) lack resources to develop dedicated platforms or tools themselves. This, in turn, makes the activists more immune to reprisals by governments, especially because shutting down a popular mainstream site could potentially backfire, drawing more attention to the cause.

Virtual Reality and Avatars

One of the first online "collaborative surprises" was the virtual world called *Second Life*. Philip Rosedale and the Linden Lab, based in San Francisco, developed Second Life in 2003 as an experimental game—"a world of collaborative construction," as Rosedale envisioned it. Second Life became increasingly popular, reaching its peak in the years 2007 to 2009.[10] From 2010 onward, its audience slowly declined to about one million regular users as of late 2014. The current exact number of users remains unknown. One of us, namely Aleksandra, spent more than two years of her real life in this virtual world, for academic as well as personal reasons. She learned Second Life's habits, communities, and rules.

From the very beginning, the user-centered non-quest format of Second Life extended to various areas and aspects: private life, social networking, education, collaboration, creative expression, politics, diplomacy, and religion.[11] It was the first large-scale social medium in which users, through their avatars, could be whoever they wanted and act without restraint. Relationships that formed in Second Life sometimes moved from the virtual to the real world, whether they were platonic or romantic. According to Wikipedia, Booperkit Moseley and Shukran Fahid were probably the first couple to meet in Second Life and then marry in real life. As Eiko Ikegami and others argue, relationships initiated in Second Life had a deeper dimension compared to those formed through other social media because avatars gave users a sense of proximity, intensifying the visual experience more than textual encounters did.[12]

The complexities of these encounters depend on whether the people behind the avatars engage in a way that is disassociative (purely for entertainment purposes), augmentative (for real-life purposes), or immersive (as if the avatars were them).[13] Through an avatar, a user can variously alter or even partly enhance a human experience in a physical, mental, or social dimension; for instance, swimming under water for hours without coming up for air. And, as William Sims Bainbridge argues in *Transhumanist Reader*, "under the right conditions *the avatar in the virtual world* can substantially affect and enhance the abilities of the user—the person who owns and operates it." The avatar demonstrates one way in which enhancement is not merely a matter of increasing the effectiveness of a person to take action, he claims, but it can also mean *an altered form of consciousness* that expands the opportunities for new social experiences and, in some cases, escape from conventional systems of moral constraints.[14] Even if some of these conclusions are far-fetched, taking into account the current stage of development of immersive technologies, there already exists a strong body of research showing that virtual beings affect our bodily and sensorimotor abilities, as well as reshape our body schematics. Many studies of first-person shooters—players who see through the eyes of their character rather than seeing from the third-person view of the characters they control—show that operating an avatar not only equips players with new affordances but also affects their memory and speed of volitional shifts in attention.[15,16] Moreover, as Michel-Ange Amorim has shown, avatars that coordinate "out-of-body" and "embodied" perspectives are relevant to social perception, that is, to understanding what another individual sees.[17]

This brings us to the issue of how realistic avatars may affect our sense of identity, social presence, being with others, and online intimacy. Sherry Turkle has argued that the roles played by a constructed identity in virtual games are an important aspect of personality development because they provide the opportunity to express unexplored parts of the self.[18] Involvement in the virtual world may invoke "second chances" for adults to resolve their identity issues, Turkle has claimed. Or the virtual experience helps to address entanglements in social relationships and to influence changes in their real-world social practices.[19] Much remains to be drawn from research in this area. For instance, Ezequiel Di Paolo and Hanne De Jaegher suggest that virtual beings in the future could provide excellent material for qualitative and neuroscientific studies of social presence and participatory sense making.[20] Researchers have already attempted to build neurophenomenological models of how we connect with others through avatars on the social level.[21] They suggest closely analyzing the experiences of virtual characters or socialized robots, seeing them as promising subjects for further studies of participatory sense making, as well as for understanding the role of imagination and mirroring behaviors in the social sphere.

These examples show that modes of being together online, particularly in very immersive environments, may have an additional cognitive and perceptive layer that is yet unexplored. Nevertheless, many Second Life users stressed that the deeper dimensions of interaction they experienced were crucial in their privates lives and heavily influenced other spheres of their activity on Second Life as well. At the peak of its popularity, many considered Second Life a "professional" platform, where people worked by performing jobs that mirrored their daily responsibilities, or in altered forms of their professions that better suited the Second Life environment. Major corporations attempted to use the online world to create and market products or services to their technology-savvy audiences. For example, IBM purchased 12 islands in Second Life for simulations of business operations. Musicians, artists, and news organizations (including CNET and Reuters) established their presence in Second Life.[22]

More importantly, though, Second Life residents constantly engaged in collaboration. Second Life grew rapidly, particularly in the initial period, because of users' efforts to create an economic zone by establishing land, real estate, and institutions based on those present in their real lives.[23,24] Many of the more immersed and involved users created virtual environments, such as semi-cosmic spaces for levitation and other ambient areas impossible to experience in reality. Virtual "goods" included everything from cars and clothes to jewelry and works of art. Second Life services were nonetheless rather limited and concentrated in the following categories: animating, building, texturing, scripting, and producing events. Apart from these formal services, though, there existed a whole realm of informal works, activities, and exchanges difficult to classify and often not monetized, because they did not resemble any typical service or product known in real life. Such unusual practices included various forms of collaborative knowledge production, such as those required to build the

Postmodern World of Pelagianism, with facilities, rules, codes of conduct, virtual infrastructures, and systems of beliefs that do not exist in the real world.

These practices also included collaborative efforts to either re-create or write new myths undertaken by the Tinies—one of the most collaborative communities of online world builders that Second Life or any other platform has ever experienced.[25] The main goal of Tinies was to build a strong community that would invert typical ways in which Second Life avatars had usually operated. Interestingly enough, even their visual forms stemmed from creative, nonconformist, and collaborative approaches to building avatars. Tinies were much smaller-than-normal avatars, and appeared in the form of cute creatures and pint-sized marvels of "virtual" imagination.

The concept of a tiny avatar stems from the innovative use of Second Life inventory. By using an animation overrider to fold the avatar's limbs in on themselves, and then covering the resulting shape with additional layers, the Second Life resident Kage Seraph made the first avatar that was much smaller than normal slider adjustments would allow (a slider is a tool in the avatar appearance menu). The concept was not entirely new. Second Life's residents had long been using animations to fit avatars into prim (in VR terms, primitive single-part) bodies, often based around a sphere or curled up into a tight ball. This was the first iteration of the idea that could walk, sit, dance, and perform other gestures required for a fully animated creation. The release of Tinies in mid-2005 was enormously successful and sparked a massive fad that spread through most of Second Life, largely based on the avatars' "cuteness factor" and novelty. Many popular Second Life locations like Neualtenburg set aside special areas for Tinies to shop and play.[26] In April 2007, for instance, Raglan Shire appeared. It was the first space of what would become a cluster of simulations catering to Tiny-centric creativity. The simulation featured a large forest city filled with various Tiny-themed wares made by many creative avatars on the grounds and platforms in the trees. Aside from offering many places to buy Tiny-themed items, Raglan Shire also began to host fun and interesting activities and events—such as Tiny sumo tournaments, art walks, musical and nonmusical performances, and football matches against other simulation communities—not just for Tinies but anyone with a Tiny-friendly disposition. A Tiny talent show featured over a dozen acts, and a music festival raised more than $100,000 LS (Linden dollars being the Second Life currency) for a Tiny charity. These examples of online collaborative endeavors reveal the importance of the Tinies: in a game like Second Life that relies on builders—the creators and developers of its land—many parts of this meticulously shaped world would not have existed without them. It has also become an inspiration for many other MMORPG-based building games (MMORPG stands for “massively multiplayer online role-playing game).

Instagram, Snapchat, and Tinder through the Lens of Collaboration

Let's leave our virtual-world exploration behind and focus on the current social media landscape.[27] Today social media have somehow abandoned the idea of providing a fully immersive online experience, even though Facebook plays with the concept through Facebook Spaces.[28] And, every now and then, we hear about new augmented reality or virtual reality social games and experiments.[29] The footage we see of ourselves online has recently replaced our immersive virtual selves, which perhaps turned out to be too cumbersome to build and required extensive emotional and resource investments. Keeping a virtual farm going or running a club in Second Life, for instance, demands almost as much effort as doing the same thing in the real world. And those tasks or responsibilities don't always bring recognition or gratification, especially because such qualities do not really operate in the Second Life realm. And yet the increasingly widespread use of facial recognition systems and emotional engagement measurements—in smartphones and other IoT devices—may change this situation, as more virtualized selves and spaces for their interactions come into play.

Nevertheless, we do find intriguing and diverse examples of platforms and apps in social media that we can use in collaborative ways and for collaborative purposes—and often differently from their designed goals and defining ideas. Such alternative uses may come about when features in existing tools are insufficient for collaborative purposes, or because of the very nature of collaboration. As we explained in the introduction, collaboration occurs wherever it can and whenever enough people are present to collaborate. In this context, let's first take a look at Instagram, and at what happened when the platform decided to structurally facilitate exchanges between users.

In the beginning, Instagram was all about capturing moments in real time, and many people still use it that way. But the platform has increasingly become a space for publicly sharing the most optimized, enchanting versions of a user's activities and life events, often in the form of professionally edited photos rather than candid snapshots. Emerging as a consequence are particular modes of exchange between users, such as *follow/unfollow*, *followback*, and *follow 4 follow*.[30] (These are similar to the reciprocity trend popular on Twitter, or the *sub 4* trend on YouTube.) Along those lines, is the basic idea behind Instagram's "shoutout" concept: two users of Instagram (often called #igers) consent to give one another a shout-out post on their own accounts, usually featuring a photo or video from the other user's feed, with instructions in the caption on how to follow that user.[31]

Although Instagram was not designed as a space for users to put other people's pictures or videos on their own accounts, a collaborative push from users has allowed that to happen. For some of the biggest Instagram accounts, with hundreds of thousands of followers, shout-outs have been a significant part of their growth strategy. Shout-outs became a decisive factor for increasing the visibility of Instagram celebrities (or influencers), for instance, and in that way also heavily boosted the popularity of the platform itself.

Whereas Instagram recommends to its users a number of other users that they might want to follow, Snapchat was and still is primarily designed for interacting with a smaller group of friends. Nonetheless, Snapchat gradually started to add more collaborative tools to allow users to post their photos and videos to custom story threads. With Custom Stories, as Snapchat calls them, users can add friends to a chosen story by selecting people from their contacts or by inviting users in a specified radius via Snapchat's geofencing feature. In the press release announcing the new story options, Snapchat claimed that they were "perfect for a trip, a birthday party, or a new baby story just for the family."[32] To create a story, users can tap the "Create Story" icon; if a story has not been updated or added to within a 24-hour period, it disappears in typical Snapchat manner.

Teams working remotely from different locales and time zones have discovered Snapchat as an internal communications tool in addition to their existing software stack. Today, when we think about professional collaborative tools for remote teams, we mainly refer to Zoom (video conferencing), Slack and Basecamp (project management apps and tools), or Hackpad (real-time collaborative text editing), to mention just a few. Although professional software stacks for internal communications are impressive, Snapchat attempts to compete by introducing a welcome, genuine human connection to relationships with other team members that surpasses an email exchange or a Skype call. Furthermore, because of the very nature of the platform and the short lifespan of its content, no one expects a high production value, which additionally fosters collaboration and inclusivity. Communication on Snapchat is instant and personal, even when used as a recruiting tool.[33,34] Although Snapchat is highly popular and boasts 100 million daily users who spend about 25 to 30 minutes on the site, it has so far avoided the kind of saturation that we see on Facebook and Twitter. Snapchat has thus created a less public medium, yet one that allows users to seamlessly maintain multiple relationships. It claims to have the personal aspect of a text message, the reach of Facebook, the visual aspect of

Instagram, and the real-time aspect of the live-streaming app Periscope. While tackling a project in teams, Snapchat allows for a gamified experience of working and sending feedback in real time. In the words of one satisfied remote team member: "You can go from 'fix it' to 'fixed' in a matter of seconds. Boom."[35]

Users also use Snapchat for documenting customer visits. When a co-founder of a company is away from headquarters or the team is geographically dispersed, founder "walks" can keep people in touch. One company describes this option in the following way: "At their own convenience, they go for a walk to go through what's on the other person's mind and snap back their replies." Apparently, many teams also use Snapchat to keep fit, which takes us back to the ideas in our chapter on collaborative gadgets. As yet another pro-Snapchat company exec boasts, "Spearheaded by Alberto "Muscles" Nodale, Close.io has turned into a team that gets excited about sweating. Through Snapchat, we send each other pep talks and gym guilt."[36]

Nevertheless, Snapchat was not meant to be an internal communications tool. The platform does not guarantee that employees will adopt it: not everyone wants to be a part of the Snapchat community, and not everyone knows how to become one. Despite some collaboration-friendly features, Snapchat mainly orients itself to reporting what has been done and does not fuel collaborative tasks, and the structural limitations of its information sharing inhibit its future potential as a communications tool. More significantly, the financial future of Snapchat and the faith in its business model remain shaky; lack of innovation, poor advertising, and competition from Instagram (owned by Facebook since 2012) contributed to plunging stock prices in fall 2018.[37,38] Given those factors, it is hard to foresee if Snapchat will become a more mainstream collaborative platform. As we write, it seems like Snapchat is struggling to find itself a viable future. On the other hand though, some of its features and functions that serve as collaboration tools (such as Stories) have been copied by other social media, ironically Instagram and Facebook.

Surprisingly, we can also consider Tinder a collaborative tool, as many do use it for purposes other than finding dating partners. One of the most interesting side functions of the app involves the "Passport" feature; when traveling solo, users can "pin" their location to a map and "swipe right" (the app's signature affirmation gesture) to connect with locals who share the same interests and might be willing to show them around; and they can do so without wanting to get involved in a romantic relationship.[39,40,41] Many Tinder-using frequent travelers meet folks who offer sightseeing tours for free, but sometimes they match up with paid professionals who provide more formal guided services. Another good reason for using Tinder while travelling is to get recommendations about popular bars and restaurants. Although the quality of online peer reviews in general is sometimes dubious,[42] some Tinder users depend on information from their "matches" to find out where to go and what to see.

One of the most unusual and yet popular uses for Tinder is political campaigning. Take the case of Robyn Gedrich, who swiped right on every Tinder profile that she encountered during Bernie Sanders's campaign in the US presidential election in 2016. When she hit a match, she would immediately send a message: "Do you feel the bern? Please text WORK to 82623 for me. Thanks!"[43] Many users confused Robyn with a bot—an issue that we consider in the last chapter of this book—but she quickly revealed that she is a 23-year-old from New Jersey with a real-life motive: to turn the Tinder dating application into a political outreach platform. And during the Polish, French, and Dutch elections from 2016 to 2018, many political parties considered Tinder to be one of their preferred platforms for grassroots mobilization. Parties like Razem, Wolność, or the less-known RiGCz in Poland set up Tinder profiles and would constantly use the "swipe right" feature to communicate with users and share updates in a more individualized or even intimate manner.[44]

Yet another collaborative use of Tinder is to acquire new language skills. A blogger who calls herself Irish Polyglot wrote in one of her posts about this fairly unexplored feature of Tinder Plus, a premium version of the app that offers unlimited swiping possibilities around the world:

> I was swiping away in London, when I "matched" with a spunky (that's Australian for handsome) boy from Prague. He was fluent in *three* languages—Czech, German and English. He lived in Austria, was learning Russian, and looking to improve his English. We were having quite a good time chatting, when he proposed moving onto Skype. I was concerned at first, imagining a situation similar to the harrowing experiences I had with Chatroulette as a teenager (anyone who stumbled onto this webpage last decade will know exactly what I'm talking about). He assured me he just wanted to talk. He called me up and we had a chat as he walked home from work. His English was far more advanced than my German, but it was quite fun. Here was an opportunity to make a new friend from a country and cultural entirely different from my own. The door was opened and Tinder immediately become a much more interesting app to use.[45]

House exchanges have become another interesting example of an "altered" use of the Tinder application and actually extend beyond the application itself.[46] In a Tinder-like manner, the Dutch HuisjeHuisje application matches rental apartments for exchange in Amsterdam, with its main goal being to revive the tight local market in social housing (affordable housing for low-income residents). If renters think about moving, they look at the apartments of others on the app, and if anything catches their fancy, they swipe right. When a match occurs between renters, they meet and discuss details so the apartment exchange can actually happen.

In these and similar examples of collaborative uses on Tinder and other platforms, how much of this collaborative turn is steered by top-down evolution of a tool that wants to grow its market share, and how much of it comes from bottom-up cases of collaborative use of the app generated by users? Platforms can track how users navigate their sites to seek out collaboration tools, and thus app manufacturers can offer innovative ways for users to market these tools or allow users to discover them on their own. This is why, despite the potentially promising collaborative dimension of social media, there is increasing concern among social media critics and some users that identifying boundaries between a community-driven collaboration and commercial exploitation will become more difficult. On the other hand, perhaps they are overreacting: seeing the issue in such a polarized way can be counterproductive, as the communality of social media use does not have to be compromised by its commercial dimension.

Does technology shape
social relations and the
collaborative spirit,
or are our needs to
collaborate so strong
that they can alter our
technology?

It may be more worthwhile to focus on whether the collaboration happens at all, and how much collaboration the users can exercise within platforms and between them. Accidental collaboration does have its shortcomings and may work well for some activities, but it proves inefficient for long-term projects that require a lot of involvement. Moreover, because this type of collaboration largely depends on functions and capabilities that are usually nonuser-generated—user-demand-driven, at best—the kinds of collaboration that the user may wish to exercise are limited and, to a large extent, managed by the platform. The issue is whether, in the future, we will see the kind of grassroots "collaborative spirit" we've already observed in peer-to-peer endeavors and wikis, but based on Tinder-like tools. If not, the collaboration practiced through applications and platforms like Tinder or Instagram will take a completely different shape. And so we ask: Does technology shape social relations and the collaborative spirit, or are our needs to collaborate so strong that they can alter our technology? This is one of the main questions that we critically analyze in the next (and final) chapter.

9

CONTROVERSIES AND THE FUTURE OF COLLABORATIVE SOCIETY

Sharing and getting information, making friends, building bonds, dating, working, campaigning, gaming—today we are doing some (or all!) of these things online. Hardware, websites, wikis, apps, and digital personas are giving new meanings to our relationships and affecting our sense of commitment and collaboration. What, though, are the emerging consequences of this digital reality for building a shared and collaborative horizon?

It seems that after decades of individualization and fragmentation, one of the biggest challenges of our age is to formulate a new definition of "we." What does "together" mean today? These questions are being addressed not only in broad public debate and major social activist movements, but also in local initiatives, such as citizens' co-ops. Our book has so far investigated ways in which the digitization of society can contribute to new forms of collectiveness. But how can this digitization play a role in the transition to a more social and, possibly, ecological society? After all, the future of collaborative society may unfold in diverse directions. Now that we have analyzed so many shades of collaboration, the important issues to address in this chapter relate to the following three questions:

(1) What essentially is and will be considered "collaborative?"

(2) With whom does collaboration count as collaboration?

(3) What are the economical and ecological costs of collaboration?

Transforming Our Collaborative Ways

To explore our first question, let's establish that despite doubts and questions concerning the sharing economy, online platforms, or the use and misuse of data, many scholars argue that the internet can still augment avenues for personal expression and promote citizen activity. As Zizi Papacharissi suggests, we may indeed be past that idyllic period when we treated internet and communication technologies (ICT) as democracy-empowering and essentially harmless; now the confluence of those technologies and the emerging Internet of Things has created a public space for politically and socially oriented conversation to flourish, addressing potential benefits as well as risks and flaws.[1,2]

When compared to people using search engines, as recent studies show, those who seek news and information from social media are at a higher risk of becoming trapped in a "collective social bubble" where news is shared within communities of like-minded individuals.[3,4] In these bubbles of filtered information, people choose to consume material that reinforces their existing attitudes. Even when opposing views appear side by side, people still select from content that upholds opinions they already agree with. The internal logic of social media platforms themselves reinforce their users' "safe" choices by offering them a narrower spectrum of sources.[5,6] In this way, online collaboration amplifies the echo chambers and further insulates the filter bubbles. For better or for worse, the process of discovery undergoes a transformation, from an individual to a social endeavor.[7]

We are not exactly clear yet how this rapidly advancing virtual layer will further translate into the real world. But there is one thing we are beginning to realize: the virtual/real public space, subjected to continuous transformation, may become more unstable, fragmented, and fragile, but it is also increasingly open for collaboration, particularly for motivated and consolidated online groups. New technologies still carry the promise of creating a sense of togetherness, even if Mark Zuckerberg's first-ever Facebook Community Summit agenda—"bringing the world closer together"—sounds less than authentic or feasible given the polarization that results from the social media giant's own part in creating filter bubbles.[8] More than ever before, society looks toward computer-mediated political communication that will actually facilitate grassroots democracy and bring people around the world together.

Observing the current digital landscape, we see that rituals of giving and sharing remain prevalent, even when it becomes difficult (or even desirable) to consider the internet as a one-size-fits-all paradigm for community-oriented communication.[9,10] We see as well that the idea of collaboration morphs in diverse ways that are not necessarily community-based, as the examples in this book demonstrate.

As Adam Arvidsson has noted in his article "Value and Virtue in the Sharing Economy," and as we have stressed in some of our previous chapters, attempting to reconcile sharing in terms of a business model that allows for economic gains, and sharing in terms of *creating commons* as a "cooperative way of doing things," poses some problems.[11] Although it is true that collaborative society thrives via the collective activities of communities and social movements, it is important to repeatedly ask what counts as collaboration.

Although it is true that collaborative society thrives via the collective activities of communities and social movements, it is important to repeatedly ask what counts as collaboration.

We have reason to doubt whether the internet space will become dominated by grassroots collaborative communities with rules and codes of conduct based on genuine communal virtues. Collaboration also exists within *disorganized crowds*, although it is then very different from the common imaginary that connects it to an *ethos* and a higher purpose.[12] Crowds, of course, have been active online for quite some time—Clay Shirky identifies a whole gamut of crowd activity in *Here Comes Everybody*—but crowds are rapidly gaining momentum.[13] Andrew McAffee and Eric Brynjoloffson, in tandem with several other authors, argue that crowds do very well in the new platform-oriented social and economic life; moreover, they are joined by artificial intelligence (AI) that is supposed to enhance their efforts and the effects of their activities.[14,15] But in specific circumstances crowds can easily change into angry mobs. *Lynching and cyberbullying can also be collaborative.* The story of the Korean "dog poop girl" is quite symptomatic. The young woman was photographed after neglecting to pick up her dog's poop from the floor of a train, even though passengers had requested she do so. Her picture, posted on a blog, went viral. As a result of the public shaming and embarrassment that followed, she dropped out of her university.[16] The story is just one example of how anonymous online collaboration has led to social stigmatizing that would likely never before, at least to such an extent, have been possible.

So, as long as tech-savvy online mobs of strangers use social media to coordinate their efforts, and rely on non-transparent algorithms to do so, the risk of abusing the system and harnessing them for the platform operator's advantage remains high.[17]

What are the alternatives to the online crowd? Some new technologies allow for platforms that intermediate free-flow communication and governance between and by users, which foster not only community building but also the emergence of *networked individualism*, a kind of internet operating system for connection, in which the person "is the focus: not the family, not the work unit, not the neighborhood, and not the social group."[18] For example, as research by Fleura Bardhi and Giana M. Eckhart on the Zipcar car-sharing network emphasizes, instead of communal relations, participants tend to foster the opposite: a "lack of identification, ... negative reciprocity resulting in a big-brother model of governance, and a deterrence of brand community."[19] But others advocate the individualization of user experience, either by reducing communication into standardized forms or by replacing the need for the collective negotiation of rules and behavior. In any case, we need to be careful with the word "community" itself. Although Zuckerberg often refers to the two billion Facebook users as a community, they have no power, responsibilities, reciprocal rights, or influence over their governance and as such are more of a surveilled, atomized user base than a community of any sort. This phenomenon is typical for many social networks and also largely detrimental to collaborative behaviors.

Networked individualism results from a particular mix of "collaborative community," and many understand it as intrinsically opposed to the logic of market exchange. As Arvidsson recalls, the kind of communities that he had observed in his research were not *collective* but rather *connective* entities. They were based not on networks of interdependence but, instead, stemmed from connected disparate projects, often in a temporary manner. As he remarks, "this makes them less susceptible to the threat of disintegration through market exchange, since there is simply much less of a communitarian substance there to disintegrate."[20] This does not mean that hopes for a digital commons should be lightly discarded, but there are various forms of control, exploitation, and bias that characterize many of the projects constructed online. As we already cautioned at the beginning of this chapter, it may also prove counterproductive to idealize collaboration.

Human-Computer Collaboration

Now we explore answers to our second question, framed in the context of AI: Who do we talk to and collaborate with on the internet? Here again we raise a legitimate question:

Who counts as part of collaborative society? Does an army of online bots also contribute to online collaboration? One of us (Aleksandra) is very much involved in creating bots and conducting various experiments with them, as she has been observing how bots and chatbots have spread in the past few years.[21,22] Currently, the largest use of bots is in web spidering or web crawling, in which an automated script fetches, analyzes, and files information from web servers at many times the speed of a human.[23,24] More than half of all web traffic comes from the activity of bots. Anyone can find plenty of instruction online about how to grow an army of bots, or can read various reports of how many such armies exist and operate in diverse online spaces.[25,26]

Today we do not think of bots as helpful computer programs but as tools for manipulation, or even as powerful weapons.[27] Spam bots bombard our email inboxes with undesired content and disturb our chats by sending unsolicited messages. Some advertisers use these bots to target individuals based on demographic and psychographic information obtained, often illegally, from user profiles. There also are zombie bots, which are compromised computer programs that become part of malicious botnets developed for large-scale attacks, in which all zombie computers act in unison, implementing commands from the master botnet owner. Those people particularly fond of online collaboration are familiar with the threat of file-sharing bots, which are the opposite of anything we would ever call beneficial collaboration. Users of peer-to-peer file sharing services usually encounter them when the bots take the users' query terms and respond by confirming that they have a file available, for which they immediately offer a link. Unsuspecting users click on the link, open the file, and unknowingly infect their computers. A current widespread version of malicious bots in the form chatbots are so advanced that they neatly emulate human interactions and obtain personal information from unsuspecting victims. Finally, there are fraud bots, running as scripts, that attempt to profit by generating false clicks for ad revenue programs, thus creating fake users for sweepstake entries and generating thousands of fake votes.

The picture is much more comprehensive. Global investigations continue to look into the extent to which citizens are influenced and exploited by advanced nudging techniques based on the combination of machine deep learning, Big Data, and sociometrics, which enable the refined profiling, microtargeting, tailoring, and manipulation of choice architectures for commercial or political purposes.[28]

Taking all that into account, it may seem fair to state that bots essentially undermine collaboration. But neither is this a complete picture. First of all, there are plenty of other uses for bots that do not compromise personalized information or content. Take art and literary bots on

Twitter, for example, such as the one that applies a quilt pattern to an image, or the one that riffs on a famous poet's verse. And second, even though bots may exacerbate the problem of fake news, their role in the field of collaboration and information sharing is, in fact, very complex. To address the worldwide concern over false news and the possibility that it can influence political, economic, and social well-being, Soroush Vosoughi and colleagues conducted a study that used a dataset of rumor cascades on Twitter between 2006 and 2017 to understand how false news spreads:

> About 126,000 rumors were spread by ~3 million people. False news reached more people than the truth; the top 1% of false news cascades diffused to between 1,000 and 100,000 people, whereas the truth rarely diffused to more than 1,000 people. Falsehood also diffused faster than the truth. The degree of novelty and the emotional reactions of recipients may be responsible for the differences observed.[29]

The most important part of the study revealed that humans are inclined to spread more false than true news, whereas bots spread both true and false information equally. This discovery contradicts conventional wisdom and provokes a few important questions. On the other hand, what if we could take a more inclusive perspective toward these new agents that carry information and communicate with humans? Then, perhaps, could we end with different criteria for web crawlers and chatbots?

Let's get technical for a moment. A chatbot system is a software program that interacts with users in "natural language" conversation, whether auditory or textual. Scholars have used different terms for chatbots, from machine conversation system to virtual agent, dialogue system, and chatterbot. Initially, developers built and used chatbots for fun. Some developers were tempted by the challenge of fooling users into thinking they were conversing with a real person. They used simple keyword-matching techniques to pair user input with a pre-programmed response that would move the conversation forward, thereby creating an illusion that the user had been understood; the first program of this kind, developed in 1966, was called ELIZA.[30,31] Before the arrival of graphical user interfaces, the 1970s and 1980s saw rapid growth in natural language interface research. Since then, a range of new chatbot architectures has appeared, such as CONVERSE, MegaHAL, and A.L.I.C.E. (artificial linguistic internet computer entity, or Alicebot). The latter stores knowledge about English conversation patterns in artificial intelligence markup language files (AIML), a derivative of the extensible markup language (XML) that Malcolm Wallace developed. The Alicebot free software community, with its highly collaborative operating manner, enabled people to input dialogue pattern knowledge into chat-based virtual agents based on the A.L.I.C.E. open-source software technology from 1995 onward. Since then a lot has changed when it comes to bot development and management. Currently, the choice of chatbots is extremely broad and includes a great deal of machine and deep learning–supported consumer technologies, such as Siri, Google Now, and Cortana, the personal digital assistants for (respectively) Apple, Goggle, and Microsoft. Some take the shape of virtual agents or physical objects, such as Amazon Alexa or the "social robot" called Jibo, and they open opportunities for research from the perspective of the proximity they maintain with the users as well as gestural communication.[32]

Currently, chatbot systems may not only mimic human conversation and entertain users, but they also figure importantly in applications for education, information retrieval, business, and most importantly e-commerce. In fact, chatbots are the perfect example of implementing state-of-the-art consumer-oriented artificial intelligence that simulates human behavior based on formal models. They have served as subjects for the research of patterns of human and nonhuman interaction, as well as issues related to assigning social roles to others, finding patterns of successful and unsuccessful interactions, and establishing social relationships and bonds.

Why shouldn't bots and chatbots develop to become more refined, nuanced, context-aware, and transparent collaboration assistants? Engineers and scientists passionate about natural language processing, generation, and understanding created chatbots primarily because people want to use their own language to communicate with computer systems in the most natural way. Janusz Kacprzyk and Slawomir Zadrozny argue that the best way to facilitate human-computer interaction is by allowing users to express their intentions, "interests, wishes, or queries directly and naturally, by speaking, typing, and pointing."[33] Kellie Morrissey and Jurek Kirakowski make a similar point in their criteria for developing of a more human-like chatbot.[34] Perhaps chatbots should become more context-aware and thus more sensitive to human needs, and thus search and deliver information that people need and would not be able to obtain otherwise. Whether we like it or not, bots and chatbots have become actors in the online world and the collaborative society. Just follow Zuckerberg's 2018 testimony at the US Senate hearings, during which the Facebook CEO explained about the influence of bots on the US presidential election in 2016. According to current and emerging market trends as well as the posthumanist and Transhumanist paradigm, it is highly likely that bots will become even more powerful interfaces of communication and agents of collaboration in the future.

Recently, MIT Media Lab's Cesar Hidalgo imagined a system of direct democracy fueled by personalized digital agents that vote on issues for the people they represent. In line with this vision, voters would be connected to digital agents that would collect information on our needs, views, and politics via the data we feed into social platforms and search engines. This, Hidalgo said, would "essentially work as a political Spotify."[35] It is surely an intriguing (although not so new) proposal—food for thought for any of us who want to examine the flaws of democratic system—but it is also potentially dangerous.

GPT-2, a deep learning–based open text generator developed by the OpenAI team, turns out to be a great example of the power and the threat posed by bots. This "large-scale unsupervised language model ... generates coherent paragraphs of text, achieves state-of-the-art performance on many language modeling benchmarks, and performs rudimentary reading comprehension, machine translation, question answering, and summarization—all without task-specific training." How was it trained? Simply to predict what the next word would be to follow a 40-gigabyte sample of internet text.[36]

The bot was actually so persuasive in how it generated stories, scientific articles, and opinion pieces that OpenAI decided to publically release a much smaller version of the model to the market, fearing it could become the text version of the "deepfakes" phenomenon in images, and thus pose a threat to social media discussions and even democracy. These examples show that more than ever before, we need to examine the hidden material contexts of all exchanges with digital technologies.

Assessing the Costs of Collaboration

Turning to our third question, many of us, all too frequently, neglect the hidden costs of ICTs and by extension the costs of our intensifying collaboration. It is now of utmost importance for a collaborative society to address them. We'll look first at the environmental crisis.[37]

Technologies that enhance collaboration come to consumers in the wealthier parts of the world as finished products. This hides the reality of costs incurred in poorer parts of the world where most of our devices and IT tools, whether desktop or mobile, are manufactured. These costs include pollution as a direct result of extracting raw materials or disposing of waste related to manufacture, as well as the emission of fossil fuels during the transport of an ever-increasing supply of goods around the world, from factory to warehouse to consumer. This relates to a phenomenon that Vasilis Kostakis and colleagues note: "while technologies can mediate abundance, they can never produce it."[38] Somewhere far away from where technology is used, someone will have to not only extract the raw material needed to ensure this superficial abundance, but also settle for less than a living wage to do so. We need to bear in mind that superefficient technologies typically encourage increased throughput of raw materials and energy. By no means do they reduce it. Data on the global use of energy and raw material in our times indicate that absolute efficiency has, in fact, never occurred. Both global energy use and global material use has increased threefold since the 1970s.[39]

Apart from ecological costs of collaboration, there are societal ones. It is hard not to notice that collaboration became both a notion and a *modus operandi* used by corporations: collaborative teams in a wide range of commercial endeavors using the Slack platform and other planning and communication tools, for instance, or software developers using Scrum's Agile Manifesto for professional development and for-profit training.[40] Just as free/open source and peer production have been successfully repurposed by capitalism, collaborative society may be re-appropriated by the economic logic of exchange. The very same platforms and technologies that currently enable it may also help to monetize it. Even the original collaborative society bonding and creation rituals, such as hackathons, are abused for profit. Sharon Zukin and Max Papadantonakis recently argued that companies use the allure of hackathons to make people work for free.[41] They observed that sponsors of hackathons fueled the "romance of digital innovation by appealing to the hackers' aspiration to be multi-dimensional agents of change" when, in fact, the hackathons were "just a means of labor control."[42]

Many of the looming risks for collaborative society development, like advancing surveillance, business-model hijacking by for-profit entities, user atomization and polarization, and labor abuses, stem from the fact that society and individual governments have failed to provide regulatory and shared ownership standards with the key stakeholders.[43] Modern technologies available at low costs and those that reap other benefits for consumers are, then, intimately connected to higher costs and burdens for workers with suppressed capacities to negotiate reasonable prices and wages. In the wealthier parts of the world, the perception of technology as magic, along with a historically unprecedented access to fossil fuels, hides the zero-sum logic that underpins an abundance enjoyed only by some.[44] We must measure the degree to which technology can alleviate scarcity, and assess it in terms of a reality in which technology exists as a strategy to bring abundance for some at the expense of others. This is as true for robots and 3D printers as it is for renewable energy technologies like solar photovoltaic modules and biofuels produced from crop residues.

In this context, as we draw our discussion of collaborative society to a close, we turn to the subject of efficiency. First we must emphasize that even though maximizing efficiency could be, in particular circumstances, considered a form of collaboration, *collaboration per se should not be mistaken for maximal efficiency*.

As we look at how other scholars are positioned on this issue, Henry Lieberman and Christopher Fry acknowledge and agree that efficiency is perhaps the most common political response to today's issues of biophysical limitations and scarcity.[45] Such a response, however, often overlooks some of the predictable consequences of efficiency that appear in profit-based economies. The Jevons Paradox, a key finding attributed to the nineteenth-century British economist Stanley Jevons, states that efficiency improvements are typically coupled with an absolute increase of consumption resulting from lower prices per unit and a subsequent increase in demand.[46] For example, the invention of more efficient steam engines allowed for cheaper transportation that catalyzed the industrial revolution, but it heavily increased the use of fossil fuels.

Ultimately, we must understand this problem in terms of rearranging resource expenditures so that efficiency improvements in one part of the world economy increase resource expenditures in the other part. Even if this occurs deep below the surface of our daily tasks and errands, efficiency always has biophysical and societal costs. Only then can we fully realize that the commoning of material knowledge and resources must lie at the heart of any technologically mediated attempt at collaboratively creating a more decent life for all, whichever form it assumes.

Collaboration will prevail. Its form, however, remains very blurry. From the late 1990s and early 2000s, we witnessed the enthusiasm toward empowerment via online communities, which extended to later disillusionment with the capacities of online mobilization; this intersected with the growing prominence of technology at the expense of human collaboration and the turbulence related to cybersafety. Recently some social media platforms acknowledged for the first time that ordinary use of their product—a euphemism for ubiquitous use—could be harmful and destructive to the social fabric, because they became mere tools for passive consumption and not necessarily for interaction.[47] Should we then dismiss the majority of the social media landscape as utterly noncollaborative? Not necessarily. We could argue that consuming information together is a particular form of collaboration, as is cyberbullying, too. It is, however, not necessarily the form of collaboration that we currently need the most. We could also try to find collaboration where it is hidden (as we did in chapter 8, "Being Together Online"), consider how to make the best out of it, and create inventory that allows collaboration to thrive. We need activity, debate, and meaningful exchanges. What is more, we have the technological tools to facilitate it. The question now is whether fostering collaboration can become a priority in terms of education, online tools creation, and the shaping of both institutional and business models in the newly emerging reality of the "post" era.

GLOSSARY

Alterscience
A collection of different, not necessarily coherent systems of convictions about the nature of the world, medicine, nutrition, health, and various other fields and topics, which are based on hearsay and beliefs rather than verified through research and the scientific methods, although not always contradicting them (e.g., Paleo diet, acupuncture).

Antiscience
Social movements and groups convinced that the scientific world is either in conspiracy with industry or simply not competent enough; they thus treat scientific knowledge with suspicion and disbelief, and actively oppose it (e.g., antivaccination movements, Flat Earthers.)

Artificial intelligence (AI)
A science and a set of computational technologies that are inspired by the ways people use their nervous systems and bodies to take actions, acquire knowledge, and reason about the world. AI is a very broad field, consisting of such domains as computer vision, natural language processing, robotic process automation, expert systems, and *machine learning*.

Biohacking
This term, broadly referring to using technology to change or influence biological organisms, can be used in diverse contexts: do-it-yourself (DIY) biology (studying biology on one's own and pursuing biological experiments); grinder biohacking (altering bodies by implanting DIY cybernetic devices and wearable technologies); nutrigenomics (using nutrition to hack human biology); an in the Quantified Self movement (for measuring activity, behaviors, and biomarkers to optimize health, well-being, and mental states).

Bot/Chatbot
Software application that runs various tasks online. On social media, bots are sets of algorithms that establish services for the users. Most famous cases are

chatbots that are able to converse in natural language and thus mimic human behavior. The history of social botting can be traced back to Alan Turing in the 1950s and his idea, known as the Turing test, of designing rules for recognizing bots and distinguishing them from humans.

Citizen science
Social movements and groups relying on scientific methods to advance or provide knowledge, organized collaboratively and outside of academic hierarchies, and allowing a wider public engagement in scholarly discovery.

Collaborative economy
See *sharing economy*

Collaborative society
An increasingly common phenomenon of emergent and enduring cooperative groups, whose members have developed particular patterns of relationships through technology-mediated cooperation.

Free/Libre open source software (F/L/OSS)
A term coined to avoid taking sides in the heated debate on whether the term "free software" or "open-source software" is preferable when referring to code that is freely licensed to any user and available to use, copy, study, modify, and reshare.

Gig economy
A phenomenon of increasing the number of occupations that rely on contingent, short-term demand, operating in radical informational asymmetry, and often enabled by intermediating online platforms.

Hacktivism
A set of online activities, with roots in hacker culture, which use technology to promote a social or political change.

Heterarchy
A form of organizing, typically used for open collaboration and peer production projects, in which the structure is nonhierarchical and any unit can be governed by others.

Human–computer interaction
The research and design of communication between computer technology and its users, mainly focused on the interface between computers and people.

Internet of Things (IoT)
The network of internet-connected devices extending beyond standard formats such as desktop computers or smartphones, to everyday objects (such as home appliances or clothing) or infrastructures (such as urban or rural utilities and transport systems). All IoT devices must contain electronics, software, actuators, monitors, or sensors, for example, which allow them to establish connectivity, exchange/receive data, and interact.

Machine learning
The study of statistical models, algorithms, and neural networks used in computer systems to improve their performance when solving various tasks. Machine learning algorithms build a model of sample data, known as a "training set," in order to make predictions or decisions on the so-called test set.

Open collaboration
A form of organization that participants can freely join or quit, and thus share a common goal, in a loosely coordinated fashion often enabled by online platforms. The phenomenon is similar to *peer production*, but with less emphasis on the sustainable outcome.

Peer production
A method of cooperating in large, self-organized communities, to produce goods or services and result in a commons-based shared outcome, often enabled by online platforms.

Platform capitalism
An economic drive within the capital market to harness the power of cutting-edge technologies so as to connect different users and turn them into workers and customers, and to arbitrate the pricing thanks to informational asymmetry.

Platform cooperativism
Using technology to connect users and allow them to exchange goods and services in a nonexploitative way.

Produsage
The process that blurs the roles of creators (active producers) and users (passive users), typically occurring in online media, F/L/OSS, or blogs.

Quantified Self
A movement as well as a practice also known as *lifelogging*: using technology (mainly *wearable technology*) for measuring activity, behaviors, and biomarkers to optimize health, well-being, and mental states.

Remix culture
This social trend for appreciating derivative art relies on the creative combining or editing existing of works, and requires the reader to be able to contextualize the original sources and denote the quotes.

Second Life
An online nonquest game developed, owned, and launched (in 2003) by the San Francisco–based company Linden Lab. It is a virtual world where users create avatars that are free to do whatever they wish, including build companies, get married, network socially, or attend events. By 2013, Second Life had approximately one million regular users. The number of current users is unknown.

Sharing economy (also called *collaborative economy*)
A very broad term related to utilizing the idle capacity of different kinds of resources and making them available for free or for profit, usually enabled by technology, and relying on internet platforms connecting strangers and intermediating trust between them. See also *unsharing economy*

Technodeterminism
A theory that assumes that a society's technology determines the development of its social, political, and economic structures as well as its cultural values. It also claims that technology is the main driver of societal change.

Unsharing economy
The term emphasizes the market-driven aspects of the so-called sharing economy, which leads to reducing the number of actual acts of genuine nonprofit sharing while accommodating and monetizing the idle capacities of different goods and services.

Wearable technology (wearable devices, tech togs, fashion electronics)
Smart electronic devices equipped with microcontrollers that can be worn on the body. Usually used for tracking various forms of users' activities and delivering data. Wearable technology is also an example or subcategory of the *Internet of Things*.

Wikipedia
The most comprehensive and cool online encyclopedia in the universe, creating the vision of a world in which every single human being can freely share in the sum of all knowledge.

NOTES

The stand-alone hyperlinks in the notes below have been shortened and time-stamped through Perma.cc.

Chapter 1

1. F. D. McMillan, "The Psychobiology of Social Pain: Evidence for a Neurocognitive Overlap with Physical Pain and Welfare Implications for Social Animals with Special Attention to the Domestic Dog (*Canis familiaris*)," *Physiology & Behavior* 167 (2016): 154–171.

2. F. Warneken, and M. Tomasello, "Helping and Cooperation at 14 Months of Age," *Infancy* 11 (2007): 271–294.

3. Y. Rekers, D. B. M. Haun, and M. Tomasello, "Children, but Not Chimpanzees, Prefer to Collaborate," *Current Biology* 21 (2011): 1756–1758.

4. F. Warneken, "How Children Solve the Two Challenges of Cooperation," *Annual Review of Psychology* 69 (2018): 205–229.

5. K. Hill, "Altruistic Cooperation during Foraging by the Ache, and the Evolved Human Predisposition to Cooperate," *Human Nature* 13 (2002): 105–128.

6. E. Fehr and U. Fischbacher, "The Nature of Human Altruism," *Nature* 425 (2003): 785–791.

7. M. D. Lieberman, *Social: Why Our Brains Are Wired to Connect* (Crown: 2013).

8. M. Nowak and R. Highfield, *SuperCooperators: Altruism, Evolution, and Why We Need Each Other to Succeed* (Simon and Schuster, 2011).

9. B. Hare, "Survival of the Friendliest: *Homo sapiens* Evolved via Selection for Prosociality," *Annual Review Psychology* 68 (2017): 155–186.

10. J. W. Van de Vondervoort and J. K. Hamlin, "The Early Emergence of Sociomoral Evaluation: Infants Prefer Prosocial Others," *Current Opinion in Psychology* 20 (2018): 77–81.

11. R. Hepach, A. Vaish, M. Tomasello, "A New Look at Children's Prosocial Motivation," *Infancy* 18 (2013): 67–90.

12. J. A. Sommerville et al., "Infants' Prosocial Behavior Is Governed by Cost-Benefit Analyses," *Cognition* 177 (2018): 12–20.

13. M. Tomasello, A. P. Melis, C. Tennie, E. Wyman, and E. Herrmann. "Two Key Steps in the Evolution of Human Cooperation: The Interdependence Hypothesis," *Current Anthropology* 53 (2012): 673–692.

14. G. Monbiot, *Out of the Wreckage: A New Politics for an Age of Crisis* (Verso Books, 2017).
15. A. Waytz and K. Gray, "Does Online Technology Make Us More or Less Sociable? A Preliminary Review and Call for Research," *Perspectives in Psychological Science* 13 (2018): 473–491.
16. R. Barbrook, *Imaginary Futures: From Thinking Machines to the Global Village* (Pluto Press, 2007).
17. Y. Takhteyev and Q. DuPont, "Retrocomputing as Preservation and Remix," *Library Hi Tech* 31 (2013): 355–370.
18. P. Himanen, *The Hacker Ethic and the Spirit of the Information Age* (Random House, 2001).
19. J. Zittrain, *The Future of the Internet and How to Stop It* (Yale University Press, 2008).
20. M. Felson and J. L. Spaeth, "Community Structure and Collaborative Consumption: 'A Routine Activity Approach.'" *American Behavioral Science* 21 (1978): 614.
21. J. Schor, "Debating the Sharing Economy," *Great Transition Initiative* (October 2014).
22. C. Shirky, *Cognitive Surplus: How Technology Makes Consumers into Collaborators* (Penguin, 2010).
23. A. Wittel, "Counter-commodification: The Economy of Contribution in the Digital Commons," *Culture and Organization* 19 (2013): 314–331.
24. Y. Benkler, "Sharing Nicely: On Shareable Goods and the Emergence of Sharing as a Modality of Economic Production," *Yale Law Journal* (2004): 273–358.
25. A. Wittel, "Qualities of Sharing and Their Transformations in the Digital Age," *International Review of Information Ethics* 15 (2011): 3–8.
26. cf. https://perma.cc/QE9Y-KK7B
27. https://perma.cc/5KQS-YFYP
28. Zittrain, *The Future of the Internet*.
29. Y. Benkler, "Practical Anarchism Peer Mutualism, Market Power, and the Fallible State," *Politics and Society* 41 (2013): 213–251.
30. R. Botsman and R. Rogers, *What's Mine Is Yours: The Rise of Collaborative Consumption* (HarperBusiness, 2010).
31. L. Gansky, *The Mesh: Why the Future of Business Is Sharing* (Penguin Press, 2010).
32. T. Terranova, *Network Culture: Politics for the Information Age* (Pluto Press, 2004).

33. E. Morozov, *The Net Delusion: The Dark Side of Internet Freedom* (Public Affairs, 2012).
34. N. A. John, *The Age of Sharing* (Polity, 2017).
35. L. Rainie and B. Wellman, *Networked: The New School Operating System* (MIT Press, 2012).

Chapter 2

1. S. S. Levine and M. J. Prietula, "Open Collaboration for Innovation: Principles and Performance," *Organization Science* 25 (2013): 1414–1433.
2. We have a slight preference for the term "open source" rather than "free" software, even though Richard Stallman fiercely opposed the former and supported the latter (see https://perma.cc/MF3R-WLE9). In his view, "open source" relates to the way software is developed, rather than the social movement and purpose conveyed by the "free software" movement. Nevertheless, currently the term may be confusing as it is occasionally hijacked by adware or freemium programs, and can be also additionally confused with programs offered for free, but harvesting user data, even though the free software creators emphasized very clearly that it is about liberties, not price. "Open source" is also roughly four times more commonly used than "free software" in Google search results.
3. https://perma.cc/T5AD-QPPE
4. Y. Benkler, *The Wealth of Networks: How Social Production Transforms Markets and Freedom* (Yale University Press, 2006).
5. D. Jemielniak, *Common Knowledge? An Ethnography of Wikipedia* (Stanford University Press, 2014).
6. For instance, consider the L'Atelier Paysan and Farmhack communities that create agricultural machines; the Wikihouse that aims at "democratizing" the construction of sustainable, resource-light dwellings; the OpenBionics project that develops designs for robotic and bionic devices; the AbilityMate that produces ankle foot orthoses; or the RepRap that creates designs for 3D printers.
7. M. M. Appleyard and H. W. Chesbrough, "The Dynamics of Open Strategy: From Adoption to Reversion." *Long Range Planning* 50 (2017): 310–321.
8. C. B. Jackson, C. Østerlund, G. Mugar, K. D. Hassman, and K. Crowston, "Motivations for Sustained Participation in Crowdsourcing: Case Studies of Citizen Science on the Role of Talk," in *48th Hawaii International Conference on System Sciences* (2015): 1624–1634.
9. Even some of the critics of for-profit peer-to-peer platforms emphasize that they rely on sharing, and consider Uber or Airbnb to be the good examples

of "sharing economy." J. Schor, "Debating the Sharing Economy," *Great Transition Initiative* (October 2014).
10. M. Kaplan, "Fijian Water in Fiji and New York: Local Politics and a Global Commodity," *Cultural Anthropology* 22 (2007): 685–706.
11. Y. Benkler, *The Wealth of Networks: How Social Production Transforms Markets and Freedom* (Yale University Press, 2006).
12. R. Botsman and R. Rogers, *What's Mine Is Yours: The Rise of Collaborative Consumption* (HarperBusiness, 2010).
13. T. Slee, *What's Yours Is Mine: Against the Sharing Economy* (OR Books, 2017).
14. N. A. John, *The Age of Sharing* (Polity, 2017).
15. Ibid.
16. P. Aigrain, *Sharing: Culture and the Economy in the Internet Age* (Amsterdam University Press, 2012).
17. N. Srnicek, *Platform Capitalism* (John Wiley & Sons, 2017).
18. V. B. Dubal, "The Drive to Precarity: A Political History of Work, Regulation, & Labor Advocacy in San Francisco's Taxi & Uber Economics," *Berkeley Journal of Employment and Labor Law* 38 (2017): 73–135.
19. K. Frenken and J. Schor, "Putting the Sharing Economy into Perspective," *Environmental Innovation and Societal Transitions* 23 (2017): 3–10.
20. Srnicek, *Platform Capitalism*, 9.
21. C. J. Martin, "The Sharing Economy: A Pathway to Sustainability or a Nightmarish Form of Neoliberal Capitalism?" *Ecological Economics* 121 (2016): 149–159.
22. L. Richardson, "Performing the Sharing Economy," *Geoforum* 67 (2015): 121–129.
23. A. Acquier, T. Daudigeos, and J. Pinkse, "Promises and Paradoxes of the Sharing Economy: An Organizing Framework," *Technological Forecasting and Social Change* 125 (2017): 1–10.
24. A. Sundararajan, *The Sharing Economy: The End of Employment and the Rise of Crowd-Based Capitalism* (MIT Press, 2016), 38.
25. A. Arvidsson, "Value and Virtue in the Sharing Economy," *Sociological Review* 66 (2018): 289–301.
26. D. Murillo, H. Buckland, and E. Val, "When the Sharing Economy Becomes Neoliberalism on Steroids: Unravelling the Controversies," *Technological Forecasting and Social Change* 125 (2017): 66–76.
27. V. B. Dubal, "The Drive to Precarity: A Political History of Work, Regulation, & Labor Advocacy in San Francisco's Taxi & Uber Economics," *Berkeley Journal of Employment and Labor Law* 38 (2017): 73–135.

28. T. Scholz, *Uberworked and Underpaid: How Workers Are Disrupting the Digital Economy* (John Wiley & Sons, 2016).
29. D. E. Sanders and P. Pattison, "Worker Characterization in a Gig Economy Viewed through an Uber Centric Lens," *Southern Law Journal* 26 (2016): 297–320.
30. A. Shapiro, "Between Autonomy and Control: Strategies of Arbitrage in the 'On-Demand' Economy," *New Media & Society* 1461444817738236 (2017).
31. J. B. Schor and W. Attwood-Charles, "The 'Sharing' Economy: Labor, Inequality, and Social Connection on For-Profit Platforms," *Sociology Compass* 11 (2017): e12493.
32. M. Bauwens and V. Kostakis, "From the Communism of Capital to Capital for the Commons: Towards an Open Co-operativism," *tripleC: Communication, Capitalism & Critique. Open Access Journal for a Global Sustainable Information Society* 12 (2014): 356–361.
33. N. van Doorn, "Platform Labor: On the Gendered and Racialized Exploitation of Low-Income Service work in the 'On-Demand' Economy," *Information, Communication and Society* 20 (2017): 898–914.
34. M. L. Gray and S. Suri, *Ghost Work: How to Stop Silicon Valley from Building a New Global Underclass* (Houghton Mifflin Harcourt, 2019).
35. N. Srnicek, *Platform Capitalism* (John Wiley & Sons, 2017).
36. Slee, *What's Yours Is Mine*.
37. I. Constantiou, A. Marton, and V. K. Tuunainen, "Four Models of Sharing Economy Platforms," *MIS Quarterly Executive* 16, no. 4 (2017): 231–251.
38. V. Kostakis, "In Defense of Digital Commoning," *Organization* 1350508417749887 (2018).
39. R. Belk, "Sharing Versus Pseudo-Sharing in Web 2.0," *The Anthropologist* 18 (2014): 7–23.
40. Srnicek, *Platform Capitalism*.
41. Scholz, *Uberworked and Underpaid*.
42. Y. Benkler and H. Nissenbaum, "Commons-Based Peer Production and Virtue," *Journal of Political Philosophy* 14 (2006): 394–419.
43. B. M. Hill and A. Monroy-Hernández, "The Cost of Collaboration for Code and Art: Evidence from a Remixing Community," in *Proceedings of the 2013 Conference on Computer Supported Cooperative Work* 1035–1046 (ACM, 2013).
44. Botsman and Rogers, *What's Mine Is Yours*.
45. W. Hu, "Taxi Medallions, Once a Safe Investment, Now Drag Owners into Debt," *New York Times*, September 10, 2017.

46. Reforge Team, "TaskRabbit's Pioneering Marketplace Model & Missed Growth Opportunities," *Reforge* (blog), October 10, 2017, https://www.reforge.com/blog/taskrabbit-marketplace-growth.

47. P. M. Lantz, E. Viruell-Fuentes, B. A. Israel, D. Softley, and R. Guzman, "Can Communities and Academia Work Together on Public Health Research? Evaluation Results from a Community-Based Participatory Research Partnership in Detroit," *Journal of Urban Health* 78 (2001): 495–507.

48. A. Brown, P. Franken, S. Bonner, N. Dolezal, and J. Moross, "Safecast: Successful Citizen-Science for Radiation Measurement and Communication after Fukushima," *Journal of Radiology Protection* 36 (2016): S82–S101.

49. A. Delfanti, "Tweaking Genes in your Garage: Biohacking between Activism and Entrepreneurship," in *Activist Media and Biopolitics: Critical Media Interventions in the Age of Biopower*, ed. W. Sützl and T. Hug (Innsbruck University Press, 2012), 163.

50. A. Diamant-Cohen and O. Golan, "Downloading Culture: Community Building in a Decentralized File-Sharing Collective," *Information, Community, & Society* 20 (2017): 1737–1755.

51. A. Wagener, "Creating Identity and Building Bridges between Cultures: The Case of 9GAG," *International Journal of Communication Systems* 8 (2014): 15.

52. Y. Benkler, "Peer Production, the Commons, and the Future of the Firm," *Strategic Organization* 15 (2017): 264–274.

53. D. Arcidiacono, A. Gandini, and I. Pais, "Sharing What? The 'Sharing Economy' in the Sociological Debate," *Sociological Review* 66 (2018): 275–288.

54. J. Kennedy, "Conceptual Boundaries of Sharing," *Information, Community, & Society* 19 (2016): 461–474.

55. E. S. Raymond, *The Cathedral and the Bazaar* (O'Reilly, 1999/2004).

56. D. Cheal, *The Gift Economy* (Routledge, 2015).

57. K. Healy, *Last Best Gifts: Altruism and the Market for Human Blood and Organs* (University of Chicago Press, 2010), 126.

Chapter 3

1. https://perma.cc/GW5R-H5ZQ

2. E. S. Raymond, *The Cathedral and the Bazaar* (O'Reilly, 1999/2004).

3. M. Bauwens and A. Pantazis, "The Ecosystem of Commons-Based Peer Production and Its Transformative Dynamics," *Sociological Review* 66 (2018): 302–319.

4. V. Kostakis and M. Bauwens, *Network Society and Future Scenarios for a Collaborative Economy* (Springer, 2014).

5. Y. Benkler, A. Shaw, and B. M. Hill, "Peer Production: A Form of Collective Intelligence," in *Handbook of Collective Intelligence*, ed. M. Michael and T. Bernstein (MIT Press, 2015).

6. V. Kostakis, K. Latoufis, M. Liarokapis, and M. Bauwens, "The Convergence of Digital Commons with Local Manufacturing from a Degrowth Perspective: Two Illustrative Cases," *Journal of Cleaner Production* 197 (2018): 1684–1693.

7. P. Konieczny, "Wikipedia: Community or Social Movement?" *Interface: A Journal for and about Social Movements* 1 (2009): 212–232.

8. G. von Krogh, S. Haefliger, S. Spaeth, and M. W. Wallin, "Carrots and Rainbows: Motivation and Social Practice in Open Source Software Development," *Mississippi Quarterly* 36 (2012): 649–676.

9. A. Sundararajan, *The Sharing Economy: The End of Employment and the Rise of Crowd-Based Capitalism* (MIT Press, 2016).

10. https://perma.cc/FQ5S-REAY

11. D. Jemielniak, *Common Knowledge? An Ethnography of Wikipedia* (Stanford University Press, 2014).

12. J. M. Reagle, *Good Faith Collaboration: The Culture of Wikipedia* (MIT Press, 2010).

13. S. O'Mahony, "The Governance of Open Source Initiatives: What Does It Mean to Be Community Managed? *Journal of Management and Governance* 11 (2007): 139–150.

14. S. Weber, *The Success of Open Source* (Harvard University Press, 2004).

15. D. Jemielniak, "Wikimedia Movement Governance: The Limits of A-hierarchical Organization," *Journal of Organizational Change Management* 29 (2016): 361–378.

16. K. R. Lakhani and E. von Hippel, "How Open Source Software Works: 'Free' User-to-User Assistance," *Research Policy* 32 (2003): 923–943.

17. O'Mahony, "The Governance of Open Source Initiatives."

18 S. Sharwood, "Buggy? Angry? LET IT ALL OUT says Linus Torvalds," *The Register* (2015), https://www.theregister.co.uk/2015/01/19/got_bugs_got_anger_just_get_them_out_says_linus_torvalds/ (accessed February 28, 2018).

19. O. Arazy, L. Yeo, and O. Nov, "Stay on the Wikipedia Task: When Task-Related Disagreements Slip into Personal and Procedural Conflicts," *Journal of the American Society for Information Science and Technology* 64 (2013): 1634–1648.

20. D. Jemielniak, "Breaking the Glass Ceiling on Wikipedia," *Feminist Review* 113 (2016): 103–108.

21. B. M. Hill and A. Shaw, "The Wikipedia Gender Gap Revisited: Characterizing Survey Response Bias with Propensity Score Estimation," *PLoS One* 8 (2013), doi:10.1371/journal.pone.0065782.
22. J. Söderberg, *Hacking Capitalism: The Free and Open Source Software Movement* (Routledge, 2015).
23. J. M. Reagle, "'Be Nice': Wikipedia Norms for Supportive Communication," *New Review of Hypermedia and Multimedia* 16 (2010): 161–180.
24. Jemielniak, *Common Knowledge? An Ethnography of Wikipedia.*
25. K. Crowston and J. Howison, "Assessing the Health of Open Source Communities," *Computer* 39 (2006): 89–91.
26. https://perma.cc/9433-8CSE
27. R. Stallman, "Viewpoint Why Open Source Misses the Point of Free Software," *Communications of the ACM* 52 (2009): 31–33.
28. Forking is a process of starting a new free/open source project with similar principles and using content and source code from the original one. It is possible because of the open licensing model used in free/open source projects. Two examples of successful forks are LibreOffice or Ubuntu.
29. M. Essex, "How Content Farms Are Hitting the E-Book Market," *Gigaom* (2011), https://gigaom.com/2011/06/23/419-the-new-market-for-e-book-spam-content/ (accessed February 27, 2018).
30. N. Tkacz, "The Politics of Forking Paths," in *Critical Point of View: A Wikipedia Reader*, ed. G. Lovink and N. Tkacz (Institute of Network Cultures, 2011).
31. D. Scaraboto, "Selling, Sharing, and Everything in Between: The Hybrid Economies of Collaborative Networks," *Journal of Consumer Research* 42 (2015): 152–176.
32. B. J. Birkinbine, "Conflict in the Commons: Towards a Political Economy of Corporate Involvement in Free and Open Source Software," *Political Economy of Communication* 2 (2015).
33. Benkler, Shaw, and Hill, "Peer Production: A Form of Collective Intelligence."
34. C. Tozzi and J. Zittrain, *For Fun and Profit: A History of the Free and Open Source Software Revolution* (MIT Press, 2017).
35. M. M. Wasko and S. Faraj, "Why Should I Share? Examining Social Capital and Knowledge Contribution in Electronic Networks of Practice," *Mississippi Quarterly* 29 (2005): 35–57.
36. Ibid.
37. C. Pentzold, "Imagining the Wikipedia Community: What Do Wikipedia Authors Mean When They Write about Their 'Community'?" *New Media & Society* 13 (2011): 704–721.

38. Y. Cai, and D. Zhu, "Reputation in an Open Source Software Community: Antecedents and Impacts," *Decision Support Systems* 91 (2016): 103–112.
39. J. Cox et al., "Doing Good Online: The Changing Relationships between Motivations, Activity and Retention among Online Volunteers," *Nonprofit and Voluntary Service Quarterly* (2017).
40. M. Haklay and P. Weber, "OpenStreetMap: User-Generated Street Maps," *IEEE Pervasive Computing* 7 (2008): 12–18.
41. T. Scholz, *Digital Labor: The Internet as Playground and Factory* (Routledge, 2012).
42. T. Terranova, "Free Labor: Producing Culture for the Digital Economy," *Social Text* 18 (2000): 33–58.
43. G. Ritzer and N. Jurgenson, "Production, Consumption, Prosumption: The Nature of Capitalism in the Age of the Digital 'Prosumer,'" *Journal of Consumer Culture* 10 (2010): 13–36.
44. N. R. Budhathoki and C. Haythornthwaite, "Motivation for Open Collaboration: Crowd and Community Models and the Case of OpenStreetMap," *American Behavioral Scientist* 57 (2012): 548–575.
45. T. Chełkowski, P. Gloor, and D. Jemielniak, "Inequalities in Open Source Software Development: Analysis of Contributor's Commits in Apache Software Foundation Projects," *PLoS One* 11 (2016), e0152976.
46. E. Ostrom, *Governing the Commons: The Evolution of Institutions for Collective Action* (Cambridge University Press, 1990).
47. D. Jemielniak, "Naturally Emerging Regulation and the Danger of Delegitimizing Conventional Leadership: Drawing on the Example of Wikipedia," in *The SAGE Handbook of Action Research*, ed. H. Bradbury (SAGE, 2015).
48. P. Konieczny, "Decision Making in the Self-Evolved Collegiate Court: Wikipedia's Arbitration Committee and Its Implications for Self-Governance and Judiciary in Cyberspace.," *International Sociology* 32 (2017): 755–774.
49. G. Mugar, "Preserving the Margins: Supporting Creativity and Resistance on Digital Participatory Platforms," *Proceedings of the ACM on Human-Computer Interaction* 1 (2017).
50. G. K. Lee and R. E. Cole, "From a Firm-Based to a Community-Based Model of Knowledge Creation: The Case of the Linux Kernel Development," *Organization Science* 14 (2003): 633–649.
51. C. Shirky, *Here Comes Everybody: The Power of Organizing without Organizations* (Penguin, 2009).
52. https://xkcd.com/386/

Chapter 4

1. A. Bruns, *Blogs, Wikipedia, Second Life, and Beyond: From Production to Produsage* (Peter Lang, 2008).

2. H. Jenkins, *Convergence Culture: Where Old and New Media Collide* (NYU Press, 2006), 24.

3. J. Van Dijck and D. Nieborg, "Wikinomics and Its Discontents: A Critical Analysis of Web 2.0 Business Manifestos," *New Media & Society* 11 (2009): 855–874.

4. D. M. Berry, *Copy, Rip, Burn: The Politics of Copyleft and Open Source* (Pluto Press, 2008).

5. E. G. Coleman and B. M. Hill, "How Free Became Open and Everything Else under the Sun," *M/C: A Journal of Media and Culture* 7 (2004).

6. L. Lessig, *The Future of Ideas: The Fate of the Commons in the Connected World* (Random House, 2001).

7. J. Zittrain, *The Future of the Internet and How to Stop It* (Yale University Press, 2008).

8. M. Castells, *The Rise of the Network Society* (Blackwell Publishers, 1996).

9. C. M. Kelty, "Theorising the Practices of Free Software: The Movement," in *Theorising Media and Practice*, ed. B. Bräuchler and J. Postill (Berghahn Books, 2010).

10. J. Löwgren and B. Reimer, *Collaborative Media: Production, Consumption, and Design Interventions* (MIT Press, 2013).

11. J. Potts et al., "Consumer Co-creation and Situated Creativity," *Industry and Innovation* 15 (2008): 459–474.

12. T. Simmonds, "Common Knowledge? The Rise of Creative Commons Licensing," *Legal Information Management* 10 (2010): 162–165.

13. S. Cunningham, "Emergent Innovation through the Coevolution of Informal and Formal Media Economies," *Television & New Media* 13 (2012): 415–430.

14. E. Boehlert, *Bloggers on the Bus: How the Internet Changed Politics and the Press* (Free Press, 2009).

15. A. Keen, *The Cult of the Amateur: How Blogs, MySpace, YouTube and the Rest of Today's User-Generated Media Are Killing Our Culture and Economy* (Nicholas Brealey, 2008).

16. Gresham's law, the economic principle that "bad money drives out good," derives from a law stated in 1519, the year Sir Thomas Gresham was born, by Nicolaus Copernicus; it is sometimes referred to as Gresham-Copernicus law.

17. L. Marshall, "'Let's Keep Music special. F—Spotify': On-Demand Streaming and the Controversy over Artist Royalties," *Creative Industries Journal* 8 (2015): 177–189.
18. C. Shirky, *Here Comes Everybody: The Power of Organizing without Organizations* (Penguin, 2009).
19. C. Hunter, D. Jemielniak, and A. Postuła, "Temporal and Spatial Shifts within Playful Work," *Journal of Organizational Change Management* 23 (2010): 87–102.
20. M. Crain, W. Poster, and M. Cherry, *Invisible Labor: Hidden Work in the Contemporary World* (University of California Press, 2016).
21. https://perma.cc/4J22-U3NC
22. J. Giles, "Internet Encyclopaedias Go Head to Head," *Nature* 438 (2005): 900–901.
23. N. J. Reavley et al., "Quality of Information Sources about Mental Disorders: A Comparison of Wikipedia with Centrally Controlled Web and Printed Sources. *Psychological Medicine* 48 (2012): 1753–1762.
24. D. Jemielniak, "Little Johnny and the Wizard of OS: The PC User as a Fool Hero," in *Organizational Olympians: Heroes and Heroines of Organizational Myths*, ed. M. Kostera (Palgrave, 2008).
25. One of the side effects of torrenting technology is that in unmodified software clients "seeding" is simultaneous to downloading ("leeching"), which means that one cannot download without sharing at least a little bit. Still, many users keep sharing after having downloaded the files.
26. J. Hamari, M. Sjöklint, and A. Ukkonen, "The Sharing Economy: Why People Participate in Collaborative Consumption," *Journal of the Association for Information Science and Technology* 67 (2016): 2047–2059.
27. A. Diamant-Cohen and O. Golan, "Downloading Culture: Community Building in a Decentralized File-Sharing Collective," *Information, Communication & Society* 20 (2017): 1737–1755.
28. C. M. Kelty, *Two Bits: The Cultural Significance of Free Software* (Duke University Press Books, 2008).
29. J. L. Beyer, *Expect Us: Online Communities and Political Mobilization* (Oxford University Press, 2014).
30. A. Bruns, *Blogs, Wikipedia, Second Life, and Beyond: From Production to Produsage* (Peter Lang, 2008).
31. L. Bennett, "Tracing Textual Poachers: Reflections on the Development of Fan studies and Digital Fandom (2014), doi:10.1386/jfs.2.1.5_1.
32. P. Booth, *A Companion to Fandom and Fan Studies* (John Wiley & Sons, 2018).

33. B. M. Hill and A. Monroy-Hernández, "The Remixing Dilemma: The Trade-Off Between Generativity and Originality," *American Behavioral Scientist* 57 (2013): 643–663.

34. L. Lessig, *Remix: Making Art and Commerce Thrive in the Hybrid Economy* (Bloomsbury Academic, 2008).

35. L. Leister, "YouTube and the Law: A Suppression of Creative Freedom in the 21st Century," *Thurgood Marshall Law Review* 37 (2011): 109.

36. E. Voigts, "Bastards and Pirates, Remixes and Multitudes: The Politics of Mash-Up Transgression and the Polyprocesses of Cultural Jazz," in *The Politics of Adaptation,* (Palgrave Macmillan, London, 2015), 82–96.

37. S. Dasgupta, W. Hale, A. Monroy-Hernández, and B. M. Hill, "Remixing as a Pathway to Computational Thinking," in *Proceedings of the 19th ACM Conference on Computer-Supported Cooperative Work & Social Computing* (ACM, 2016). 1438–1449.

38. A. Delwiche and J. J. Henderson, *The Participatory Cultures Handbook* (Routledge, 2012).

39. R. Botsman and R. Rogers, *What's Mine Is Yours: The Rise of Collaborative Consumption* (HarperBusiness, 2010).

40. J. van Dijck, "Users Like You? Theorizing Agency in User-Generated Content," *Media, Culture & Society* 31 (2009): 41–58.

41. https://perma.cc/2VUW-V7TN

42. L. Shifman, *Memes in Digital Culture* (MIT Press, 2014).

43. https://perma.cc/M7EG-CYAX

44. https://perma.cc/28SD-XJXK

45. C. A. Rentschler and S. C. Thrift, "Doing Feminism in the Network: Networked Laughter and the 'Binders Full of Women' Meme," *Feminist Theory* 16 (2015): 329–359.

46. A. Nagle, *Kill All Normies: Online Culture Wars from 4chan and Tumblr to Trump and the Alt-Right* (John Hunt Publishing, 2017).

47. M. S. Bernstein et al., "4chan and/b: An Analysis of Anonymity and Ephemerality in a Large Online Community," *International Conference on Weblogs and Social Media* (2011): 50–57.

48. W. L. Bennett and A. Segerberg, "The Logic of Connective Action: Digital Media and the Personalization of Contentious Politics," *Information, Communication & Society* 15 (2012): 739–768.

49. https://perma.cc/WRJ4-ACNB

50. Milner, R. M. "Pop Polyvocality: Internet Memes, Public Participation, and the Occupy Wall Street Movement," *International Journal of Communication Systems* 7, no. 34 (2013).

51. H. E. Huntington, "Pepper Spray Cop and the American Dream: Using Synecdoche and Metaphor to Unlock Internet Memes' Visual Political Rhetoric," *Communication Studies* 67 (2016): 77–93.
52. L. Silvestri, "Mortars and Memes: Participating in Pop Culture from a War Zone," *Media, War & Conflict* 9 (2015): 27–42.

Chapter 5

1. S. Vegh, M. D. Ayers, and M. McCaughey, "Classifying Forms of Online Activism," *Cyberactivism: Online Activism in Theory and Practice* (2003): 71–95.
2. P. Dourish and G. Bell, *Divining a Digital Future: Mess and Mythology in Ubiquitous Computing* (MIT Press, 2011).
3. N. Watson, "Community: A Case Study of the Phish.Net Fan Community," *Virtual Culture: Identity and Communication in Cybersociety* 102 (1997).
4. Ibid.
5. Y. Benkler, "A Free Irresponsible Press: WikiLeaks and the Battle over the Soul of the Networked Fourth Estate," *Harvard Civil Rights–Civil Liberties Law Review* 46 (2011): 311.
6. E. G. Coleman, *Coding Freedom* (Princeton University Press, 2013).
7. Benkler, "A Free Irresponsible Press."
8. L. A. Lievrouw, "WikiLeaks | WikiLeaks and the Shifting Terrain of Knowledge Authority," *International Journal of Communication Systems* 8 (2014): 2631–2645.
9. https://perma.cc/WHW8-LXHQ
10. C. Kirsch, "The Grey Hat Hacker: Reconciling Cyberspace Reality and the Law," *Northern Kentucky Law Review* 41 (2014): 383.
11. B. Schneier, *Secrets and Lies: Digital Security in a Networked World* (John Wiley & Sons, 2011).
12. G. Coleman, *Hacker, Hoaxer, Whistleblower, Spy: The Many Faces of Anonymous* (Verso, 2014).
13. N. Klein, "Culture Jamming: Ads under Attack," *Brandweek* 41 (2000): 28–35.
14. *The Blackwell Dictionary of Modern Social Thought*, ed. W. Outhwaite and W. Outhwaite, (Blackwell Publishing, 2002), vi–vi.
15. LLC Books, ed. *Cult of the Dead Cow: Hacktivismo, Demon Roach Underground, Hacktivismo Enhanced-Source Software License Agreement, Hohocon* (General Books, 2010).
16. J. A. Farmer, "The Spector of Crypto-Anarchy: Regulating Anonymity-Protecting Peer-to-Peer Networks," *Fordham Law Review* 72 (2003): 725.
17. https://perma.cc/L4C6–7ZFK

18. L. C. Keith, "The United Nations International Covenant on Civil and Political Rights: Does It Make a Difference in Human Rights Behavior?" *Journal of Peace Research* 36 (1999): 95–118.
19. R. Deibert, J. Palfrey, R. Rohozinski, J. Zittrain, and J. G. Stein, *Access Denied: The Practice and Policy of Global Internet Filtering* (MIT Press, 2008).
20. M. S. Bernstein et al., "4chan and/b: An Analysis of Anonymity and Ephemerality in a Large Online Community," *International Conference on Weblogs and Social Media* (2011), 50–57.
21. 4chan, *Daring Do and the Jungle of Terror* (CreateSpace Independent Publishing Platform, 2012).
22. J. Call, "A is for Anarchy, V is for Vendetta: Images of Guy Fawkes and the Creation of Postmodern Anarchism," *Anarchist Studies* (2008).
23. L. Goode, "Anonymous and the Political Ethos of Hacktivism," *Popular Communication* 13 (2015): 74–86.
24. N. Hampson, "Hacktivism, Anonymous & a New Breed of Protest in a Networked World," *Boston College International & Comparative Law Review* 1 (2011): 1–33.
25. A. Kobayashi, *New Type of Social Network in Business Organization: "swarm Ba"* (California State University, Sacramento, 2004).
26. P. A. Gloor, *Swarm Leadership and the Collective Mind: Using Collaborative Innovation Networks to Build a Better Business* (Emerald Publishing, 2017).
27. M. Casadei and M. Viroli, "Applying Self-Organizing Coordination to Emergent Tuple Organization in Distributed Networks," *2008 Second IEEE International Conference on Self-Adaptive and Self-Organizing Systems* (2008): 213–222.
28. Q. Norton, "Anonymous 101: Introduction to the Lulz," *Wired*, November 8, 2011.
29. G. Coleman, "Anonymous in Context: The Politics and Power behind the Mask," paper no. 3, *Internet Governance Series* (Centre for International Governance Innovation, 2013).
30. G. A. Yukl, *Leadership in Organizations* (Pearson Education India, 2013).
31. M. J. Warren, "Hackers and Cyber Terrorists," in *Encyclopedia of Information Ethics and Security*, ed. M. Quigley (IGI Global, 2007), 8.
32. G. Lucas, *Ethics and Cyber Warfare: The Quest for Responsible Security in the Age of Digital Warfare* (Oxford University Press, 2016).
33. T. Raizer, "Corruption: Luxembourg progresse, avant LuxLeaks," *Paperjam* (2014).

34. B. Huesecken and M. Overesch, "Tax Avoidance through Advance Tax Rulings—Evidence from the LuxLeaks Firms" (2015), doi:10.2139/ssrn.2664631.
35. E. Mollick, "The Dynamics of Crowdfunding: An Exploratory Study," *Journal of Business Venturing* 29 (2014): 1–16.
36. M. A. Doan and M. Toledano, "Beyond Organization-Centred Public Relations: Collective Action through a Civic Crowdfunding Campaign," *Public Relations Review* 44 (2018): 37–46.
37. Z. Tufekci, *Twitter and Tear Gas: The Power and Fragility of Networked Protest* (Yale University Press, 2017).
38. H. S. Christensen, "Political Activities on the Internet: Slacktivism or Political Participation by Other Means?" *First Monday* 16 (2011).
39. D. Karpf, "Online Political Mobilization from the Advocacy Group's Perspective: Looking Beyond Clicktivism," *Policy & Internet* 2 (2010): 7–41.
40. M. Halupka, "Clicktivism: A Systematic Heuristic," *Policy & Internet* (2014).
41. M. White, *The End of Protest: A New Playbook for Revolution* (Knopf Canada, 2016).
42. R. Nader, *The Good Fight: Declare Your Independence and Close the Democracy Gap* (Harper Collins, 2004).
43. E. Zuckerman, "New Media, New Civics?" *Policy & Internet* 6 (2014): 151–168.
44. Coleman, *Coding Freedom.*
45. Coleman, *Hacker, Hoaxer, Whistleblower, Spy.*
46. J. Nazario, "Politically Motivated Denial of Service Attacks," *The Virtual Battlefield: Perspectives on Cyber Warfare* (2009): 163–181.
47. R. Amin, "The Empire Strikes Back: Social Media Uprisings and the Future of Cyber Activism," *Kennedy School Review* 10 (2009): 64.
48. J. Scott and D. Spaniel, *Hacking Elections Is Easy!: Preserving Democracy in the Digital Age* (CreateSpace Independent Publishing Platform, 2016).
49. L. Goode, "Anonymous and the Political Ethos of Hacktivism," *Popular Communication* 13 (2015): 74–86.
50. G. Alberts and R. Oldenziel, eds., *Hacking Europe: From Computer Cultures to Demoscenes* (Springer, 2014).
51. N. Thompson et al., "Inside the Two Years That Shook Facebook—and the World," *Wired* February 12, 2018.
52. "More Snowden Leaks Reveal Hacking by NSA and GCHQ against Communications Firm." *Network Security* (2015): 1–2.
53. https://perma.cc/HXA4-XVKU

54. T. Jordan and P. Taylor, *Hacktivism and Cyberwars* (Routledge, 2004).
55. P. S. Ryan, "War, Peace, or Stalemate: Wargames, Wardialing, Wardriving, and the Emerging Market for Hacker Ethics," *Virginia Journal of Law and Technology* 9 (2004): 41.
56. P. Himanen, *The Hacker Ethic* (Random House, 2010).

Chapter 6

1. D. Innerarity, *The Democracy of Knowledge* (Bloomsbury Publishing USA, 2013).
2. I. B. Cohen, "Commentary: The Fear and Distrust of Science in Historical Perspective," *Science, Technology, & Human Values* 6 (1981): 20–24.
3. S. Rider and M. A. Peters, *Post-Truth, Fake News: Viral Modernity & Higher Education* 49 (2018): 563–566.
4. T. L. Cooper, Book Review: T. Nichols, *The Death of Expertise: The Campaign Against Established Knowledge and Why It Matters* (Oxford University Press, 2017) in *Public Administration Review* 78 (2018): 318–320.
5. P. Dahlgren, "Media, Knowledge and Trust: The Deepening Epistemic Crisis of Democracy," *Javnost—The Public* 25 (2018): 20–27.
6. K. Camargo Jr., and R. Grant, "Public Health, Science, and Policy Debate: Being Right Is Not Enough," *American Journal of Public Health* 105 (2015): 232–235.
7. M. Bijlefeld and S. K. Zoumbaris, *Encyclopedia of Diet Fads: Understanding Science and Society, 2nd Edition: Understanding Science and Society* (ABC-CLIO, 2014).
8. S. Rautiainen, J. E. Manson, A. H. Lichtenstein, and H. D. Sesso, "Dietary Supplements and Disease Prevention—A Global Overview," *Nature Reviews Endocrinology* 12 (2016): 407–420.
9. P. Mirowski, *Science-Mart* (Harvard University Press, 2011).
10. D. M. J. Lazer et al., "The Science of Fake News," *Science* 359 (2018): 1094–1096.
11. National Science Board, *Science and Engineering Indicators 2012* (National Science Foundation, 2012).
12. European Commission, "Use of the Internet," *Digital Agenda for Europe* (2012), https://ec.europa.eu/digital–agenda/sites/digital–agenda/files/scoreboard_life_online.pdf.
13. D. Brossard and D. A. Scheufele, "Science, New Media, and the Public," *Science* 339 (2013): 40–41.
14. S. Vosoughi, D. Roy, and S. Aral, "The Spread of True and False News Online," *Science* 359 (2018): 1146–1151.

15. B. W. Hesse, D. E. Nelson, G. L. Kreps et al., "Trust and Sources of Health Information: The Impact of the Internet and Its Implications for Health Care Providers: Findings from the First Health Information National Trends Survey," *Archives of Internal Medicine* 165 (2005): 2618–2624.
16. R. Dragusin et al., "Rare Disease Diagnosis as an Information Retrieval Task," *Advances in Information Retrieval Theory* 6931 (2011): 356–359.
17. M. G. Bouwman, Q. G. A. Teunissen, F. A. Wijburg, and G. E. Linthorst, "'Doctor Google' Ending the Diagnostic Odyssey in Lysosomal Storage Disorders: Parents Using Internet Search Engines as an Efficient Diagnostic Strategy in Rare Diseases," *Archives of Disease in Childhood* 95 (2010): 642–644.
18. W. E. Bijker, E. Bal, and R. Hendriks, *The Paradox of Scientific Authority: The Role of Scientific Advice in Democracies* (MIT Press, 2009).
19. D. M. Bowman, N. Woodbury, and E. Fisher, "Decoupling Knowledge and Expertise in Personalized Medicine: Who Will Fill the Gap?" *Expert Review of Precision Medicine and Drug Development* 1 (2016): 345–347.
20. F. Godlee, J. Smith, and H. Marcovitch, "Wakefield's Article Linking MMR Vaccine and Autism was Fraudulent," *British Medical Journal* 342 (2011), c7452.
21. M. Callon, and V. Rabeharisoa, "Research 'in the Wild' and the Shaping of New Social Identities," *Technology in Society* 25 (2003): 193–204.
22. S. Epstein, "The Construction of Lay Expertise: AIDS Activism and the Forging of Credibility in the Reform of Clinical Trials," *Science, Technology, & Human Values* 20 (1995): 408–437.
23. D. Durant, "Public Participation in the Making of Science Policy," *Perspectives in Science* 18 (2010): 189–225.
24. D. Weinberger, *Too Big to Know: Rethinking Knowledge Now That the Facts Aren't the Facts, Experts Are Everywhere, and the Smartest Person in the Room Is the Room* (Basic Books, 2014).
25. M. Zaród, "Stabilised Instability. Hacking Tournament as a Laboratory," Working Paper, *Science, Technology, & Human Values* (2018).
26. H. Ledford, "Biohackers Gear Up for Genome Editing," *Nature* 524 (2015): 398–399.
27. C. Jefferson, F. Lentzos, and C. Marris, "Synthetic Biology and Biosecurity: Challenging the 'Myths,'" *Frontiers in Public Health* 2 (2014): 115.
28. A. Wexler, "The Practices of Do-It-Yourself Brain Stimulation: Implications for Ethical Considerations and Regulatory Proposals," *Journal of Medical Ethics* 42 (2016): 211–215.
29. P. Rabinow and G. Bennett, *Designing Human Practices: An Experiment with Synthetic Biology* (University of Chicago Press, 2012).

30. M. Meyer, "The Politics and Poetics of DIY Biology," in *Biotechnologies, Synthetic Biology, A Life and the Arts*, ed. A. Bureaud, R. Malina, and L. Whiteley (MIT Press, 2014).
31. F. Ramella and C. Manzo, "Into the Crisis: Fab Labs—A European Story," *Sociology Review* 66 (2018): 341–364.
32. M. Meyer, "Domesticating and Democratizing Science: A Geography of Do-It-Yourself Biology," *Journal of Material Culture* 18 (2013): 117–134.
33. M. E. Vargo, *The Weaponizing of Biology: Bioterrorism, Biocrime and Biohacking* (McFarland, 2017).
34. E. Callaway, "Glowing Plants Spark Debate," *Nature* 498 (2013): 15–16.
35. Jefferson, Lentzos, and Marris, "Synthetic Biology and Biosecurity."
36. Rabinow and Bennett, *Designing Human Practices*.
37. C. Marris, C. Jefferson, and F. Lentzos, "Negotiating the Dynamics of Uncomfortable Knowledge: The Case of Dual Use and Synthetic Biology," *Biosocieties* 9 (2014): 393–420.
38. H. Wolinsky, "The FBI and Biohackers: An Unusual Relationship: The FBI Has Had Some Success Reaching Out to the DIY Biology Community in the USA, but European Biohackers ...," *EMBO Reports (*2016).
39. V. Kostakis, V. Niaros, and C. Giotitsas, "Production and Governance in Hackerspaces: A Manifestation of Commons-Based Peer Production in the Physical Realm?" *International Journal of Cultural Studies* 18 (2014): 555–573.
40. L. Irani, "Hackathons and the Making of Entrepreneurial Citizenship," *Science, Technology, & Human Values* 40 (2015): 799–824.
41. A. Brown, P. Franken, S. Bonner, N. Dolezal, and J. Moross, "Safecast: Successful Citizen-Science for Radiation Measurement and Communication after Fukushima." *Journal of Radiological Protection* 36 (2016): S82–S101.
42. A. Lis and A. K. Stasik, "Hybrid Forums, Knowledge Deficits and the Multiple Uncertainties of Resource Extraction: Negotiating the Local Governance of Shale Gas in Poland," *Energy Research & Social Science* 28 (2017): 29–36.
43. D. Jemielniak and D. J. Greenwood, "Wake Up or Perish: Neo-Liberalism, the Social Sciences, and Salvaging the Public University," *Cultural Studies↔ Critical Methodologies* 15 (2015): 72–82.
44. D. McQuillan, "The Countercultural Potential of Citizen Science," *M/C Journal* 17 (2014).
45. P. Vallabh, H. Lotz-Sisitka, R. O'Donoghue, and I. Schudel, "Mapping Epistemic Cultures and Learning Potential of Participants in Citizen Science Projects," *Conservation Biology* 30 (2016): 540–549.

46. B. Salter, Y. Zhou, and S. Datta, "Hegemony in the Marketplace of Biomedical Innovation: Consumer Demand and Stem Cell Science," *Social Science & Medicine* 131 (2015): 156–163.
47. P. Mirowski, "Against Citizen Science," *Aeon* (2017), https://aeon.co/essays/is-grassroots-citizen-science-a-front-for-big-business.
48. S. Visvanathan, "Alternative Science," *Theory, Culture & Society* 23 (2006): 164–169.
49. R. J. Bernstein, *Beyond Objectivism and Relativism: Science, Hermeneutics, and Praxis* (University of Pennsylvania Press, 2011).
50. M. Goldner, "The Dynamic Interplay between Western Medicine and the Complementary and Alternative Medicine Movement: How Activists Perceive a Range of Responses from Physicians and Hospitals," *Sociology of Health and Illness* 26 (2004): 710–736.
51. D. A. Broniatowski et al., "Weaponized Health Communication: Twitter Bots and Russian Trolls Amplify the Vaccine Debate," *American Journal of Public Health* 108 (2018): 1378–1384.
52. Z. Bauman, *Consuming Life* (Cambridge: Polity, 2007).
53. E. E. Pariser, *The Filter Bubble: What the Internet Is Hiding from You* (Penguin Press, 2011).
54. B. N. Duarte, "Entangled Agencies: New Individual Practices of Human-Technology Hybridism through Body Hacking," *Nanoethics* 8 (2014): 275–285.
55. G. Seyfried, L. Pei, and M. Schmidt, "European Do-It-Yourself (DIY) Biology: Beyond the Hope, Hype and Horror," *Bioessays* 36 (2014): 548–551.
56. R. Bolton and R. Thomas, "Biohackers: The Science, Politics, and Economics of Synthetic Biology," *Innovations* 9 (2014): 213–219 (2014).
57. R. Sennett, *The Craftsman* (Yale University Press, 2008).

Chapter 7

1. Of course, oral poetry still exists in many forms, including poetry slams, live rap, and readings. It is, however, less prominent in the highly industrialized West and its cultural role is different from and less prominent than in Homeric times.
2. R. Godwin-Jones, "Blogs and Wikis: Environments for Online Collaboration," *Language and Learning Technologies* 7, no. 2 (2003) 12–16.
3. D. R. Garrison, "Online Collaboration Principles," *Journal of Asynchronous Learning Networks* 10 (2006): 25–34.
4. J. A. West and M. L. West, *Using Wikis for Online Collaboration: The Power of the Read-Write Web* (John Wiley & Sons, 2009).

5. L. Lomicka and G. Lord, *The Next Generation: Social Networking and Online Collaboration in Foreign Language Learning* (Computer Assisted Language Instruction Consortium, 2009).
6. D. Wright, S. Gutwirth, M. Friedewald, E. Vildjiounaite, and Y. Punie, *Safeguards in a World of Ambient Intelligence* (Springer, 2008).
7. D. Kozlov, J. Veijalainen, and Y. Ali, "Security and Privacy Threats in IoT Architectures," in *Proceedings of the 7th International Conference on Body Area Networks* (Institute for Computer Sciences, Social-Informatics and Telecommunications Engineering, 2012), 256–262.
8. B. Henne, C. Szongott, and M. Smith, "SnapMe If You Can: Privacy Threats of Other Peoples' Geo-tagged Media and What We Can Do About It," in *Proceedings of the Sixth ACM Conference on Security and Privacy in Wireless and Mobile Networks* (ACM, 2013), 95–106.
9. C. Cederström and A. Spicer, *The Wellness Syndrome* (John Wiley & Sons, 2015).
10. B. Bloomfield and K. Dale, "Fit for Work? Redefining 'Normal' and 'Extreme' through Human Enhancement Technologies," *Organization* 22 (2015): 552–569.
11. S. J. Thompson, *Global Issues and Ethical Considerations in Human Enhancement Technologies* (IGI Global, 2014).
12. M. Swan, "Sensor Mania! The Internet of Things, Wearable Computing, Objective Metrics, and the Quantified Self 2.0," *Journal of Sensor and Actuator Networks* 1 (2012): 217–253.
13. R. A. Calvo and D. Peters, *Positive Computing: Technology for Wellbeing and Human Potential* (MIT Press, 2014).
14. R. Wright and L. Keith, "Wearable Technology: If the Tech Fits, Wear It," *Journal of Electronic Resources in Medical Libraries* 11, (2014): 204–216.
15. F. da Costa and F. de Sá-Soares, "Authenticity Challenges of Wearable Technologies," *Managing Security Issues and the Hidden Dangers of Wearable Technologies* 33 (2016).
16. E. M. Guizzo, "The Essential Message: Claude Shannon and the Making of Information Theory (MIT Press, 2003).
17. M. Swan, "Sensor Mania! The Internet of Things, Wearable Computing, Objective Metrics, and the Quantified Self 2.0," *Journal of Sensor and Actuator Networks* 1 (2012): 217–253.
18. T. O. Zander and C. Kothe, "Towards Passive Brain–Computer Interfaces: Applying Brain–Computer Interface Technology to Human–Machine Systems in General," *Journal of Neural Engineering* 8 (2011): 025005.
19. Wright and Keith, "Wearable Technology."

20. R. Burrows and M. Savage, "After the Crisis? Big Data and the Methodological Challenges of Empirical Sociology," *Big Data & Society* 1 (2014): 2053951714540280.
21. E. A. Rake et al., "Personalized Consent Flow in Contemporary Data Sharing for Medical Research: A Viewpoint," *Biomed Research International* (2017): 7147212.
22. T. Schiphorst, "Breath, Skin and Clothing: Using Wearable Technologies as an Interface into Ourselves," *International Journal of Performance Arts and Digital Media* 2 (2006): 171–186.
23. A. C. Norris, *Essentials of Telemedicine and Telecare* (John Wiley & Sons, 2001).
24. G. Burg, *Telemedicine and Teledermatology* (Karger Medical and Scientific Publishers, 2003).
25. Y. Erlich and A. Narayanan, "Routes for Breaching and Protecting Genetic Privacy," *Nature Reviews Genetic* 15 (2014): 409–421.
26. https://perma.cc/7KTH-JFLG
27. R. Cooter, "Biocitizenship," *Lancet* 372 (2008): 1725.
28. M. Swan, "Health 2050: The Realization of Personalized Medicine through Crowdsourcing, the Quantified Self, and the Participatory Biocitizen," *Journal of Personal Medicine* 2 (2012): 93–118.
29. M. Foucault, *Society Must Be Defended: Lectures at the Collège de France, 1975–76*, trans. David Macey (Picador, 2003).
30. D. Lupton, *The Quantified Self* (John Wiley & Sons, 2016).
31. https://perma.cc/FY5V-KTBA
32. https://perma.cc/K5EJ-JFLD
33. https://perma.cc/WRE6-RQV6
34. https://perma.cc/GP5L-4R94
35. E. Zelkha, B. Epstein, S. Birrell, and C. Dodsworth, "From Devices to Ambient Intelligence," *Digital Living Room Conference* 6 (1998).
36. H. Achten, "Book Review: Digital Ground: Architecture, Pervasive Computing, and Environmental Knowledge," *International Journal of Architectural Computing* 3 (2005): 255–258.
37. J. Wu, H. Li, S. Cheng, and Z. Lin, "The Promising Future of Healthcare Services: When Big Data Analytics Meets Wearable Technology," *Information & Management* 53 (2016): 1020–1033.
38. "Wearables and the IoT for Healthcare," in *Connected Health*, ed. R. Krohn, D. Metcalf, and P. Salber, (CRC Press, 2017), 1–6.
39. https://perma.cc/F8KT-56U2; K. C. Montgomery, J. Chester, and K. Kopp, *Health Wearable Devices in the Big Data Era: Ensuring Privacy,*

Security, and Consumer Protection, Center for Digital Democracy, School of Communication, American University, Washington, DC, 2016, https://www.democraticmedia.org/sites/default/files/field/public/2016/aucdd_wearablesreport_final121516.pdf.
40. Cederström and Spicer, *The Wellness Syndrome*.
41. Swan, "Health 2050."
42. C. J. Heyes, "Foucault Goes to Weight Watchers," *Hypatia* 21 (2006): 126–149.
43. E. L. Murnane et al., "Self-Monitoring Practices, Attitudes, and Needs of Individuals with Bipolar Disorder: Implications for the Design of Technologies to Manage Mental Health," *Journal of the American Medical Informatics Association* 23 (2016): 477–484.
44. A. R. Jadad, M. Fandiño, R. and Lennox, "Intelligent Glasses, Watches and Vests ... Oh My! Rethinking the Meaning of 'Harm' in the Age of Wearable Technologies," *Journal of the American Medical Informatics Association Journal of Medical Internet Research mHealth uHealth* 3 (2015): e6.
45. D. Lupton, "Quantified Sex: A Critical Analysis of Sexual and Reproductive Self-Tracking Using Apps," *Culture, Health & Sexuality* 17 (2015): 440–453.
46. https://perma.cc/UD26-FJBA
47. https://perma.cc/3LWM-PVWN
48. https://perma.cc/4PGB-67AF
49. E. Shove and M. Pantzar, "Consumers, Producers and Practices Understanding the Invention and Reinvention of Nordic Walking," *Journal of Consumer Culture* 5 (2005): 43–64.
50. Lupton, *The Quantified Self*.
51. D. Nafus and K. Tracey, "13 Mobile Phone Consumption and Concepts of Personhood," *Perpetual Contact* 206 (2002).
52. D. Nafus and J. Sherman, "Big Data, Big Questions | This One Does Not Go up to 11: The Quantified Self Movement as an Alternative Big Data Practice," *International Journal of Communication Systems* 8 (2014): 11.
53. S. Hales, C. Dunn, S. Wilcox, and G. M. Turner-McGrievy, "Is a Picture Worth a Thousand Words? Few Evidence-Based Features of Dietary Interventions Included in Photo Diet Tracking Mobile Apps for Weight Loss," *Journal of Diabetes Science Technology* 10 (2016): 1399–1405.
54. S. S. Gorin, *Prevention Practice in Primary Care* (Oxford University Press, 2014).
55. C. Thompson-Felty and C. S. Johnston, "Adherence to Diet Applications Using a Smartphone Was Associated With Weight Loss in Healthy Overweight

Adults Irrespective of the Application," *Journal of Diabetes Science Technology* 11 (2017): 184–185.
56. E. G. Coleman, *Coding Freedom* (Princeton University Press, 2013).
57. L. Breiman, J. Friedman, C. J. Stone, and R. A. Olshen, *Classification and Regression Trees* (Routledge, 2017).
58. Cited in https://perma.cc/7MFJ-WTQH; see also D. L. Johnson, *The Chilean Road to Socialism* (Anchor, 1973).
59. https://perma.cc/7MFJ-WTQH
60. https://perma.cc/UW64-DRL3

Chapter 8

1. C. Chung and K. Austria, "Social Media Gratification and Attitude toward Social Media Marketing Messages: A Study of the Effect of Social Media Marketing Messages on Online Shopping Value," *Proceedings of the Northeast Business & Economics Association* (2010).
2. R. C. M. Maia, "Media, Social Change, and the Dynamics of Recognition," in *Recognition and the Media*, ed. R. C. M. Maia (Palgrave Macmillan UK, 2014) 181–198.
3. R. S. Tokunaga and J. D. Quick, "Impressions on Social Networking Sites: Examining the Influence of Frequency of Status Updates and Likes on Judgments of Observers," *Media Psychology* (2017): 1–25.
4. S. Vandoninck, L. d'Haenens, R. De Cock, and V. Donoso, "Social Networking Sites and Contact Risks among Flemish Youth," *Childhood* 19 (2011): 69–85.
5. D. Boyd, "Why Youth (Heart) Social Network Sites: The Role of Networked Publics in Teenage Social Life," in *Youth, Identity, and Digital Media*, ed. D. Buckingham (MIT Press, 2007).
6. A. Lenhart and M. Madden, *Teens, Privacy & Online Social Networks: How Teens Manage Their Online Identities and Personal Information in the Age of MySpace* (Pew Internet & American Life Project, 2007).
7. D. Boyd, "Can Social Network Sites Enable Political Action?" *International Journal of Media & Cultural Politics* 4 (2008): 241–244.
8. J. G. Breslin, A. Passant, and S. Decker, "Introduction to the Social Web (Web 2.0, Social Media, Social Software)," in *The Social Semantic Web*, ed. J. G. Breslin, A. Passant, S. and Decker, (Springer Berlin Heidelberg, 2009), 21–44.
9. E. Zuckerman, "Cute Cats to the Rescue" (chapter 6), in *From Voice to Influence: Understanding Citizenship in a Digital Age*, ed. D. Allen and J. S. Light (University of Chicago Press, 2015).
10. Rosedale quoted in S. Totilo, "Do-It-Yourselfers Buy into This Virtual World," *New York Times*, November 11, 2004; W. Evans, "Hungry Ghosts

in Second Life," in *Information Dynamics in Virtual Worlds* (Elsevier, 2011), 97–104.

11. Second Life is, of course, not the only virtual space where various bonds were formed and communities shaped. It is, however, a great example of the first nonquest games that focused on building social relations and collaboration.

12. E. Ikegami and P. Hut, "Avatars Are for Real: Virtual Communities and Public Spheres," *Journal of Virtual Worlds Research* (2008).

13. A. Przegalinska, "Embodiment, Engagement and the Strength Virtual Communities: Avatars of Second Life in Decay," *Tamara Journal for Critical Organization Inquiry* 13 (2015).

14. W. S. Bainbridge, "Transavatars," chapter 7, in *The Transhumanist Reader: Classical and Contemporary Essays on the Science, Technology, and Philosophy of the Human Future*, ed. M. More and N. Vita-More (John Wiley & Sons, 2013), our emphasis.

15. A. D. H. Teng, N. C. H. Chee, and C. Y. K. Goh, "Apparatus, System, and Method for Providing Independent Multi-screen Viewing," *US Patent* (2013).

16. A. R. Galloway, *Gaming: Essays on Algorithmic Culture* (University of Minnesota Press, 2006), 18.

17. M.-A. Amorim, "'What Is My Avatar Seeing?': The Coordination of 'Out-of-Body' and 'Embodied' Perspectives for Scene Recognition Across Views," *Visual Cognition* 10 (2003): 157–199.

18. S. Turkle, *The Second Self: Computers and the Human Spirit* (MIT Press, 2005).

19. S. Turkle, "Constructions and Reconstructions of Self in Virtual Reality: Playing in the MUDs," *Mind, Culture, and Activity* 1 (1994): 158–167.

20. H. De Jaegher and E. Di Paolo, "Participatory Sense-Making," *Phenomenological Cognitive Science* 6 (2007): 485–507.

21. D. Cressman and E. Hamilton, "The Experiential Dimension on Online Learning: Phenomenology, Technology and Breakdowns," *Glimpse* 7 (2005): 9–21.

22. A. Bruns, *Blogs, Wikipedia, Second Life, and Beyond: From Production to Produsage* (Peter Lang, 2008).

23. A. Przegalinska, "Embodiment, Engagement and The Strength Virtual Communities."

24. A. K. Przegalinska, *Istoty wirtualne* (Universitas, 2016).

25. Bruns, *Blogs, Wikipedia, Second Life, and Beyond*.

26. https://perma.cc/L7SH-EB3V

27. T. Kelsey, *Social Networking Spaces: From Facebook to Twitter and Everything in Between* (Apress, 2010).
28. R. Marfone, *The Real Reason Facebook Acquired Oculus Rift: How Virtual Reality Will Disrupt Every Industry and Why You Should Care* (CreateSpace Independent Publishing Platform, 2016).
29. P. A. Smith, "Using Commercial Virtual Reality Games to Prototype Serious Games and Applications," in *Virtual, Augmented and Mixed Reality*, ed. S. Lackey and J. Chen, 10280, (Springer International Publishing, 2017), 359–368.
30. Y. Hu, L. Manikonda, S. Kambhampati et al., "What We Instagram: A First Analysis of Instagram Photo Content and User Types," in *Proceeding of the 8th Annual International Conference on Weblogs and Social Media* (AAII Press, 2014), 595–598.
31. J. Phua, S. V, Jin, and J. J. Kim, "Gratifications of Using Facebook, Twitter, Instagram, or Snapchat to Follow Brands: The Moderating Effect of Social Comparison, Trust, Tie Strength, and Network Homophily on Brand Identification, Brand Engagement, Brand Commitment, and Membership Intention," *Telematics and Informatics* 34 (2017): 412–424.
32. https://perma.cc/AJR8-MTT2
33. R. K. Logan, "Social Media Including Twitter, Instagram, and Snapchat," in *Understanding New Media* (Peter Lang, 2016).
34. R. Basler, *Snapchat im Recruiting: Was Wir von Social Media fuers HR lernen koennen* (CreateSpace Independent Publishing Platform, 2016).
35. https://perma.cc/Q75R-QYJP
36. https://perma.cc/K8F3-F6XS
37. https://perma.cc/355R-Z67C
38. https://perma.cc/LX9B-C3FC; https://perma.cc/6DJ2–5JDL; https://perma.cc/AJ2X-XV7G
39. S. R. Sumter, L. Vandenbosch, and L. Ligtenberg, "Love Me Tinder: Untangling Emerging Adults' Motivations for Using the Dating Application Tinder," *Telematics and Informatics* 34 (2017): 67–78.
40. S. Bouwmeester, "Keep Playing: Tinder, Its Affordances and Playful Identity," Bachelor thesis, University of Utrecht Repository (2016).
41. T. H. Nazer, F. Morstatter, G. Tyson, H. Liu, "A Close Look at Tinder Bots," https://pdfs.semanticscholar.org/c846/c8166fa113df9c55bf7b9875b9de97225dc8.pdf.
42. https://perma.cc/D2XW-QXYW
43. https://perma.cc/VLC2-HEYK; https://perma.cc/38FK-D5ZV
44. https://perma.cc/Z49T-8P2T

45. https://perma.cc/QG3G-6RGN
46. https://perma.cc/6982-MPF9

Chapter 9

1. Z. Papacharissi, "The Virtual Sphere: The Internet as a Public Sphere," *New Media & Society* 4 (2002): 9–27.
2. For a comprehensive picture of IoT, see: J. Gubbi, R. Buyya, S. Marusic, and M. Palaniswami, "Internet of Things (IoT): A Vision, Architectural Elements, and Future Directions," *Future Generation Computing Systems* 29 (2013/9): 1645–1660.
3. A. M. Kaplan and M. Haenlein, "Users of the World, Unite! The Challenges and Opportunities of Social Media," *Business Horizons* 53 (2010): 59–68.
4. C. Fuchs, *Social Media: A Critical Introduction* (Sage, 2017).
5. S. Jean Tsang, "Cognitive Discrepancy, Dissonance, and Selective Exposure," *Media Psychology* (2017): 1–24.
6. J. Choi and J. K. Lee, "Investigating the Effects of News Sharing and Political Interest on Social Media Network Heterogeneity," *Computers in Human Behavior* 44 (2015): 258–266.
7. D. Nikolov, D. F. M. Oliveira, A. Flammini, and F. Menczer, "Measuring Online Social Bubbles," *PeerJ Computer Science* 1 (2015): e38.
8. https://perma.cc/4KY5-8MPJ
9. C. Marín Beristain, D. Paez, and J. L. González, "Rituals, Social Sharing, Silence, Emotions and Collective Memory Claims in the Case of the Guatemalan Genocide," *Psicothema* 12 (2000).
10. A. P. Fiske, *Structures of Social Life: The Four Elementary Forms of Human Relations: Communal Sharing, Authority Ranking, Equality Matching, Market Pricing* (Free Press, 1991).
11. A. Arvidsson, "Value and Virtue in the Sharing Economy," *Sociology Review* 66 (2018): 289–301.
12. U. Dolata and J.-F. Schrape, "Masses, Crowds, Communities, Movements: Collective Action in the Internet Age," *Social Movement Studies* 15 (2016): 1–18.
13. C. Shirky, *Here Comes Everybody: The Power of Organizing without Organizations* (Penguin, 2008).
14. D. G. Goldstein, R. P. McAfee, and S. Suri, "The Wisdom of Smaller, Smarter Crowds," in *Proceedings of the Fifteenth ACM Conference on Economics and Computation* (ACM, 2014), 471–488.
15. A. McAfee and E. Brynjolfsson, *Machine, Platform, Crowd: Harnessing Our Digital Future* (W. W. Norton & Company, 2017).

16. D. J. Solove, *The Future of Reputation: Gossip, Rumor, and Privacy on the Internet* (Yale University Press, 2007).
17. H. Rheingold, *Smart Mobs: The Next Social Revolution* (Basic Books, 2002).
18. L. Rainie and B. Wellman, *Networked: The New Social Operating System* (MIT Press, 2012), 6.
19. F. Bardhi and G. M. Eckhardt, "Access-Based Consumption: The Case of Car Sharing," *Journal of Consumer Research* 39, no 4 (2012): 881–898.
20 Arvidsson, "Value and Virtue in the Sharing Economy."
21. L. Ciechanowski, A. Przegalinska, M. Magnuski, and P. Gloor, "In the Shades of the Uncanny Valley: An Experimental Study of Human–Chatbot Interaction," *Future Generations of Computing Systems* (2018). doi:10.1016/j.future.2018.01.055.
22. L. Ciechanowski, A. Przegalinska, and K. Wegner, "The Necessity of New Paradigms in Measuring Human-Chatbot Interaction," in *Advances in Cross-Cultural Decision Making*, ed. M. Hoffman (Springer International Publishing, 2018), 205–214.
23. S. Zheng, *Effective Methods for Web Crawling and Web Information Extraction* (Pennsylvania State University, 2011).
24. C. Olston and M. Najork, "Web Crawling," *Foundations and Trends in Information Retrieval* 4 (2010): 175–246.
25. https://trafficleaks.com/bot-army
26. http://blogs.discovermagazine.com/d-brief/2017/01/20/twitter-bot-army/#.XBrD5xNKjOQ
27. https://perma.cc/5RCJ-D7RZ
28. https://perma.cc/CZ47–4BF9
29. S. Vosoughi, D. Roy, and S. Aral, "The Spread of True and False News Online," *Science* 359 (2018): 1146–1151.
30. J. Weizenbaum, "ELIZA—A Computer Program for the Study of Natural Language Communication between Man and Machine," *Communications of the ACM* 9 (1966): 36–45.
31. P. Auslander, "Live from Cyberspace: or, I Was Sitting at My Computer This Guy Appeared He Thought I Was a Bot," *PAJ: A Journal of Performance and Art* 24 (2002): 16–21.
32. Ciechanowski, Przegalinska, Magnuski, and Gloor, "In the Shades of the Uncanny Valley."
33. J. Kacprzyk and S. Zadrozny, "Computing with Words Is an Implementable Paradigm: Fuzzy Queries, Linguistic Data Summaries, and Natural-Language Generation," *IEEE Transactions on Fuzzy Systems* 18 (2010): 461–472.

34. K. Morrissey and J. Kirakowski, "'Realness' in Chatbots: Establishing Quantifiable Criteria," in *International Conference on Human-Computer Interaction* (Springer, 2013), 87–96.
35. https://www.fastcompany.com/40557688/this-plan-for-an-ai-based-direct-democracy-outsources-votes-to-a-predictive-algorithm
36. https://perma.cc/FC7Z-E53X
37. V. Kostakis, K. Latoufis, M. Liarokapis, and M. Bauwens, "The Convergence of Digital Commons with Local Manufacturing from a Degrowth Perspective: Two Illustrative Cases," *Journal of Cleaner Production* (2016), doi:10.1016/j.jclepro.2016.09.077.
38. Ibid.
39. https://perma.cc/24FX-ULKK
40. C. C. Heckscher, P. S. Adler, and P. Adler, *The Firm as a Collaborative Community: Reconstructing Trust in the Knowledge Economy* (Oxford University Press, 2007).
41. S. Zukin and M. Papadantonakis, "Hackathons as Co-optation Ritual: Socializing Workers and Institutionalizing Innovation in the 'New' Economy," in *Precarious Work*, 157–181.
42. https://perma.cc/V7E8-LT2W
43. N. Schneider, "An Internet of Ownership: Democratic Design for the Online Economy," *Sociology Review* 66 no. 9 (2018): 320–340.
44. https://perma.cc/TDF5-C2AK
45. H. Lieberman and C. Fry, "Will Software Ever Work?" *Communications of the ACM* 44 (2001): 122–124.
46. https://perma.cc/8VFF-S25A
47. https://perma.cc/S3CR-7U5D

FURTHER READING

More about collaborative consumption

Botsman, R., and R. Rogers. *What's Mine Is Yours: The Rise of Collaborative Consumption.* New York: HarperBusiness, 2014.

Kostakis, V., and M. Bauwens. *Network Society and Future Scenarios for a Collaborative Economy*. New York: Springer, 2014.

Lessig, L. *Remix: Making Art and Commerce Thrive in the Hybrid Economy.* New York: Penguin, 2008.

More about digital copyright and open access

Haber, E. *Criminal Copyright*. Cambridge: Cambridge University Press, 2018.

Lessig, L. *Free Culture: How Big Media Uses Technology and the Law to Lock Down Culture and Control Creativity*. New York: Penguin, 2004.

Suber, P. *Knowledge Unbound: Selected Writings on Open Access, 2002–2011*. Cambridge, MA: MIT Press, 2016.

More about digital platforms and sustainability

Gray, M. L., and S. Suri. *Ghost Work: How to Stop Silicon Valley from Building a New Global Underclass*. Boston: Houghton Mifflin Harcourt, 2019.

Rainie, L., and B. Wellman. *Networked: The New Social Operating System*. Cambridge, MA: MIT Press, 2012.

Srnicek, N. *Platform Capitalism*. New York: John Wiley & Sons, 2017.

Terranova, T. *Network Culture: Cultural Politics for the Information Age*. New York: Pluto Press, 2004.

More about the internet culture

Castells, M. *Communication Power*. Oxford: Oxford University Press, 2013.

Raymond, E. S. *The Cathedral and the Bazaar*. Cambridge, MA: O'Reilly, 1999/2004.

Shifman, L. *Memes in Digital Culture*. Cambridge, MA: MIT Press, 2014.

Zittrain, J. *The Future of the Internet and How to Stop It*. New Haven: Yale University Press, 2008.

More about online activism and inequality

Alper, M. *Giving Voice: Mobile Communication, Disability, and Inequality*. Cambridge, MA: MIT Press, 2017.

Coleman, G. *Hacker, Hoaxer, Whistleblower, Spy: The Many Faces of Anonymous*. New York: Verso Books, 2014.

Neblo, M. A., K. M. Esterling, and D. M. Lazer. *Politics with the People: Building a Directly Representative Democracy*. Cambridge: Cambridge University Press, 2018.

Tufekci, Z. *Twitter and Tear Gas: The Power and Fragility of Networked Protest*. New Haven: Yale University Press, 2017.

More about online relationships

Bruns, A. *Blogs, Wikipedia, Second Life, and Beyond: From Production to Produsage*. New York: Peter Lang, 2008.

More, M., and N. Vita-More. *The Transhumanist Reader: Classical and Contemporary Essays on the Science, Technology, and Philosophy of the Human Future*. New York: John Wiley & Sons, 2013.

Solove, D. J. *The Future of Reputation: Gossip, Rumor, and Privacy on the Internet*. New Haven: Yale University Press, 2007.

Turkle, S. *The Second Self: Computers and the Human Spirit*. Cambridge, MA: MIT Press, 2005.

More about peer production

Benkler, Y. *The Wealth of Networks: How Social Production Transforms Markets and Freedom*. New Haven: Yale University Press, 2006.

Shirky, C. *Here Comes Everybody: The Power of Organizing without Organizations*. New York: Penguin, 2006.

Surowiecki, J. *The Wisdom of Crowds*. New York: Anchor Books, 2005.

More about post-truth, fake news, and pseudoscience

Benkler, Y., R. Faris, and J. Roberts. *Network Propaganda: Manipulation, Disinformation, and Radicalization in American Politics*. Oxford: Oxford University Press, 2018.

Hecht, D. K., E. Lobato, C. Zimmerman, F. Blanco, H. Matute, D. K. Simonton, K. M. Folta, B. Hermes, L. Kennair, and E. Sandseter. *Pseudoscience: The Conspiracy Against Science*. Cambridge, MA: MIT Press, 2018.

McIntyre, L. *Post-truth*. Cambridge, MA: MIT Press, 2018.

More about sharing culture

Aigrain, P. *Sharing: Culture and the Economy in the Internet Age*. Amsterdam: Amsterdam University Press, 2012.

John, N. A. *The Age of Sharing*. Cambridge, UK: Polity, 2017.

Sundararajan, A. *The Sharing Economy: The End of Employment and the Rise of Crowd-based Capitalism*. Cambridge, MA: MIT Press, 2016.

More about studying the digital society

Gloor, P. A. *Swarm Creativity: Competitive Advantage through Collaborative Innovation Networks*. Oxford: Oxford University Press, 2006.

Hine, C. *Ethnography for the Internet: Embedded, Embodied and Everyday*. London: Bloomsbury Publishing, 2015.

Jemielniak, D. *Thick Big Data: Doing Digital Social Sciences*. Oxford: Oxford University Press, 2020.

Pavón, J., M. Arroyo, S. Hassan, and C. Sansores. "Agent-Based Modelling and Simulation for the Analysis of Social Patterns." *Pattern Recognition Letters* 29, no. 8 (2008): 1039–1048.

More about wearables, tracking, and cybersecurity

Neff, G., and D. Nafus. *Self-Tracking*. Cambridge, MA: MIT Press, 2016.

Przegalinska, A. *Wearable Technologies in Organizations: Privacy, Efficiency and Autonomy in Work*. London: Palgrave, 2019.

Schneier, B. *Click Here to Kill Everybody: Security and Survival in a Hyperconnected World*. New York: W. W. Norton & Co., 2018.

More about Wikipedia

Jemielniak, D. *Common Knowledge? An Ethnography of Wikipedia*. Stanford, CA: Stanford University Press, 2014.

Lih, A. *The Wikipedia Revolution: How a Bunch of Nobodies Created the World's Greatest Encyclopedia*. New York: Hyperion, 2009.

Reagle, J. M. *Good Faith Collaboration: The Culture of Wikipedia*. Cambridge, MA: MIT Press, 2010.

Tkacz, N. *Wikipedia and the Politics of Openness*. Chicago: University of Chicago Press, 2015.

INDEX

The MIT Press Essential Knowledge Series

AI Ethics, Mark Coeckelbergh
Auctions, Timothy P. Hubbard and Harry J. Paarsch
The Book, Amaranth Borsuk
Carbon Capture, Howard J. Herzog
Citizenship, Dimitry Kochenov
Cloud Computing, Nayan B. Ruparelia
Collaborative Society, Dariusz Jemielniak and Aleksandra Przegalinska
Computational Thinking, Peter J. Denning and Matti Tedre
Computing: A Concise History, Paul E. Ceruzzi
The Conscious Mind, Zoltan E. Torey
Contraception, Donna Drucker
Critical Thinking, Jonathan Haber
Crowdsourcing, Daren C. Brabham
Cynicism, Ansgar Allen
Data Science, John D. Kelleher and Brendan Tierney
Deep Learning, John D. Kelleher
Extraterrestrials, Wade Roush
Extremism, J. M. Berger
Fake Photos, Hany Farid
fMRI, Peter A. Bandettini
Food, Fabio Parasecoli
Free Will, Mark Balaguer
The Future, Nick Montfort
GPS, Paul E. Ceruzzi
Haptics, Lynette A. Jones
Information and Society, Michael Buckland
Information and the Modern Corporation, James W. Cortada
Intellectual Property Strategy, John Palfrey
The Internet of Things, Samuel Greengard
Irony and Sarcasm, Roger Kreuz
Machine Learning: The New AI, Ethem Alpaydin
Machine Translation, Thierry Poibeau
Macroeconomics, Felipe Larraín B.
Memes in Digital Culture, Limor Shifman
Metadata, Jeffrey Pomerantz
The Mind–Body Problem, Jonathan Westphal
MOOCs, Jonathan Haber

Neuroplasticity, Moheb Costandi
Nihilism, Nolen Gertz
Open Access, Peter Suber
Paradox, Margaret Cuonzo
Post-Truth, Lee McIntyre
Quantum Entanglement, Jed Brody
Recycling, Finn Arne Jørgensen
Robots, John Jordan
School Choice, David R. Garcia
Self-Tracking, Gina Neff and Dawn Nafus
Sexual Consent, Milena Popova
Smart Cities, Germaine R. Halegoua
Spaceflight, Michael J. Neufeld
Spatial Computing, Shashi Shekhar and Pamela Vold
Sustainability, Kent E. Portney
Synesthesia, Richard E. Cytowic
The Technological Singularity, Murray Shanahan
3D Printing, John Jordan
Understanding Beliefs, Nils J. Nilsson
Virtual Reality, Samuel Greengard
Waves, Frederic Raichlen

DARIUSZ JEMIELNIAK is Professor of Management at Kozminski University, Poland, where he heads the Management in Networked and Digital Societies (MINDS) Department, and the author of *Common Knowledge? An Ethnography of Wikipedia*. He is Faculty Associate at the Berkman-Klein Center for Internet and Society at Harvard University, is a Research Fellow at MIT's Center for Collective Intelligence, and serves on the Wikimedia Foundation Board of Trustees.

ALEKSANDRA PRZEGALINSKA is Assistant Professor at Kozminski University and was recently a Research Fellow at MIT's Center for Collective Intelligence. She is the author of *Wearable Technologies in Organizations: Privacy, Efficiency and Autonomy in Work*.